CLASSICS
IN SCIENCE

CLASSICS
IN SCIENCE

A Course of Selected Reading
by Authorities

★

Particular attention is directed to the
Introductory Reading Guide by
E. N. Da C. ANDRADE, F.R.S.

KENNIKAT PRESS
Port Washington, N. Y./London

CLASSICS IN SCIENCE

Copyright, 1960, Cultural Publications (I. U. S.), Ltd.
Reissued in 1971 by Kennikat Press by arrangement
with Philosophical Library, Inc.
Library of Congress Catalog Card No: 71-122975
ISBN 0-8046-1356-7

Manufactured by Taylor Publishing Company   Dallas, Texas

ESSAY AND GENERAL LITERATURE INDEX REPRINT SERIES

CONTENTS

vi

CONTENTS

BOOK IV

SCIENCE AND EVERYDAY LIFE

viii CONTENTS

INTRODUCTORY READING GUIDE

BY

PROF. E. N. DA C. ANDRADE, F.R.S., D.Sc., PH.D., HON. LL.D.

WHAT IS SCIENCE?

A SENSE of design and order, and a delight in classification and system, is characteristic of civilized man. In all branches of art—music, painting and all the plastic arts, poetry, architecture—we see a deliberate grouping and ordering, a putting of certain things together as a balanced whole. The justification for the way in which the arrangement is carried out, for the method according to which the notes are arranged in a sonata, for instance, must be sought in some innate character of the mind, in something that may be called an aesthetic sense, for want of a more precise term. The artist in his arrangements and juxtapositions is concerned with subjective things, with personal feelings, beliefs, sensations, and inner convictions, and nobody can bring anything approaching a proof that he is a wrong-minded or a bad artist. The only appeal is to canons of taste, to the accepted opinions of practitioners and critics, which may be widely divided, while even the verdict of the majority will vary from age to age, and from clime to clime.

Science does not differ from art in being an ordering and arrangement, nor, in spite of widespread belief, is the man of science deprived of the aesthetic satisfaction that a well-balanced piece of work gives the artist. He can often with justice speak of a beautiful experiment, even if all that is observed is the movement of a spot of light through a measured distance, of an elegant deduction, and of a pretty proof. It is in its subject-matter and in its criteria of truth that science differs from art and, for that matter, from philosophy, as the word is understood nowadays.

The subject-matter of science is the body of observed facts of Nature, and the test of the truth of a deduction is its agreement with experiment or observation. In art, the masterpieces of one generation may be derided as worthless by the next : in philosophy, a beautiful chain of reasoning can claim merit even when it makes no appeal to any such hard facts as are the subject of general agreement. In science, however, the achievements of one generation represent something won from Nature, which remains as definite gain and definite progress : an experiment properly carried out remains for all time. The most ingenious argument has no validity if it leads to a result which is contradicted by experiment. The sense impressions of the external world are the ultimate court of appeal, and the innate sense of the fitness of things, which may vary from person to person, and from school to school, has to yield to observation of what lies outside the observer. There is in science no book or canon to which all difficulties can be referred, except the book of Nature. There is no appeal to the emotion nor invocation of agencies lying outside laws deduced from the experience of many independent observers.

The Dawn of Science. This has not always been so. In Greek times, for instance, when the art of pure reasoning was carried to as high a degree of perfection as ever since, a mystic element was freely invoked in attempts to explain Nature. The school of Pythagoras sought for a connexion between the lengths of strings which gave out related musical notes and the distances of the planets from the earth, according to some divine harmony—it was in this connexion that the phrase " the music of the spheres " arose. Ten was the perfect number, and so celestial perfection demanded that there should be ten revolving bodies which, when the earth, moon, sun, and sphere of the stars had been added to the five planets, still demanded the addition of one more, the invisible " counter-earth ".[1] A mysterious perfection was also attributed to circular motion.

In the middle ages an ascription of human characteristics to inanimate Nature was general, a common mystery being believed to govern individual man and the universe : moral conceptions were applied to material phenomena. Thus the alchemists taught that metals were generated in the earth from seeds, and that they could be killed, mortified and corrupted, and born again in an incorruptible form. They had a material part and a spiritual part, and the various chemical processes were described in terms of the material and spiritual life of man. In a sense we may say that they were governed by the idea of a unity of Nature so profound that the teachings of the Church and of the philosophers were of universal application, governing both man and minerals. The chemist was one who had specialized in studying the morals of metals. Man, the microcosm, was in himself the epitome and pattern of the macrocosm, that is, of the universe at large, a general belief of which astrology, which traced the individual history of a man to the

---

[1] The " counter-earth " was demanded for other reasons as well ; the point is that this type of argument was allowed validity.

influences of the stars and other heavenly bodies prevailing at his birth, was but one aspect.

We find even in Galileo's time such contentions as these : *What are the parts of the microcosm, man, through which he receives air and enlighten- ment ? Two nostrils, two eyes, two ears, and a mouth. So in heaven, the macrocosm, there must be seven planets.* The nature of the planets is further explained in terms of the parts of the microcosm, and the argument is strengthened by the fact that there are seven metals. Hence Galileo's claim to have discovered with his telescope satellites round Jupiter was disallowed : the fact that he had seen them was a trifling circumstance unworthy of the serious attention of philosophers. Like- wise there could be no spots in the sun, as Galileo contended on the basis of his observations, because the sun was the eye of the world, and the eye of the world could not be subject to imperfections. The men who brought forward this type of objection were far from being fools : they were, or could have made, acute lawyers and theologians. Arguing from their own premises, they were capable of ingenious trains of thought : their attitude was not so much stupid as unscientific. The object of citing these things is not to deride worthy and learned men, but to show how modern a creation is the scientific attitude. The authority of experiment and of observation of the external world, as we know it, is a growth of the last three hundred years.

Aristotle. Brief as must be our mention of the historical aspect, the question of the birth of modern science can scarcely be left without a reference to Aristotle. The great Greek philosopher was a man who himself made many acute observations on the actual behaviour of plants and animals, but he left a body of false doctrine in physical science which was taken as infallible by the mediaevalists and their immediate successors. He taught, for instance, that the speed of a falling body was proportionate to its weight, and even Galileo, who first established the true law of fall, only gradually won free from the Aristotelian conceptions. Aristotle himself was a friend of experiment and observa- tion, but his authority in the sixteenth century was such that, if he had pronounced on a subject, experiment was considered to be impertinent. Even in 1668, just after the Royal Society had been founded,[1] Joseph Glanvil found it necessary to write a book defending the new plan of experiment against one who had contended " that Aristotle had more advantages for knowledge than the Royal Society or all the present age had or could have ", and who scoffed at the idea that discoveries of worth could be made with the telescope. This was within a few years

---

[1] It was founded in 1660 and its first Royal Charter granted in 1662. The founda- tion of the Royal Society was a great event in the history of science.

of Newton's great discovery that the motions of the heavenly bodies
could be explained by gravitational attraction : that the same force
which caused a body to fall to the earth, diminished by distance according to the inverse square law, tended to cause the moon to fall towards
the earth, and that this force was just counterbalanced by the outward
acceleration due to the moon's circular path round the earth.

Newton. It may perhaps be claimed that this discovery of Newton's
is the most significant recorded in the history of science. Kepler had
discovered the laws of planetary motion, but while he had found the
true mathematical shape of the orbits and the rate at which the planet
described them, he had no idea of any physical law to account for their
moving in this way, and, still under the influence of the old school of
thought, had felt obliged to suppose that there was a moving spirit,
an *anima motrix*, which drove them on. Newton showed that the
properties of matter, as known to us on this earth, were sufficient to
account for the motions of the planets, and revealed the possibilities
of describing all the motions of the external world in terms of a few
simple laws. He also expressly disclaimed any concern with what lay
behind these laws—*hypotheses non fingo*, I do not frame hypotheses, he
said, by which he did not mean that he made no hypotheses, for the
law of gravitation is a hypothesis, but that he made none that were not
susceptible to calculation and to comparison with observation, no hypothesis of a fanciful or mysterious kind. " I do not deal in conjectures "
he said, and again, of the laws of Nature, " their Truth appearing to us
by Phaenomena, though their Causes be not yet discovered." This
abstention from inquiring into what lies behind scientific laws—for
scientific laws give us the how, but not the ultimate why—is characteristic of modern science.

Facts of Nature. I may seem to have insisted too much, perhaps,
upon the fact that science is concerned with external Nature, and that
all her laws are suggested and checked by observed fact. Obvious,
however, as it may appear to be, it is not infrequently overlooked. The
theory of relativity, for instance, is often popularly discussed as if it
were a matter of pure thinking, or could, at any rate, be argued from
the ordinary facts of mechanics familiar to a schoolboy. Actually the
theory was propounded to give a logical way out of a difficulty raised
by a series of complicated experiments, of which that of Michelson and
Morley is the best known and most conclusive. It is, namely, a fact
that no terrestrial experiments have ever succeeded in revealing any
such effect of the earth's motion through space as would be anticipated
on the old theory of a fixed luminiferous ether. This result is incor-
porated in the first, the so-called special, theory of relativity. The

experiment of Michelson and Morley has been often repeated ; if a positive effect could be incontrovertibly detected and agreed by scientific men, then the theory would have to be abandoned or modified. It is, and will remain, a logical and self-consistent scheme in any case. At present we believe that it represents the facts of Nature, but it is not beyond conjecture that this may not be so, although at present all the evidence is in its favour.

Another not infrequent example of popular failure to grasp the essential nature of science is the misuse of scientific words to lend a spurious appearance of knowledge to the unknown. This is no new thing, as the familiar example from Molière's *Malade imaginaire* suffices to show. In the course of a satirical representation of the examination for the Doctor's degree, the candidate is asked, in grotesque dog Latin, why opium induces sleep, and replies that it is its dormitive virtue that causes this effect. Many who will deride such idle use of words to produce an appearance of science will, however, gravely discuss whether thought waves may not constitute an explanation of telepathy. The term " waves ", however, is used in science to describe a form of propagation of energy with perfectly definite, experimentally established, characteristics—a finite velocity of travel, and the properties of reflection, refraction, and diffraction, which are explained in the text-books. To talk of waves where no transmission of energy has been proved, and where none of their properties have been experimentally observed, is to play with words, and to ignore the character of scientific thought.

THE PHYSICAL SCIENCES

Field of Physics. Turning to the more particular tasks of science, we will in the first case confine ourselves to the physical sciences, leaving mathematics, which occupies a peculiar place, and the sciences of living matter for subsequent mention. Physics concerns itself with the phenomena shown by inanimate matter in which there is no change in the state of combination of the matter, the study of the inter-combination of different kinds of matter constituting chemistry. It deals, then, with the most fundamental properties of matter, and as the more we know of these properties, the more we can understand such particular manifestations as chemical combinations, physics is all the time extending its domain into regions which were before appropriated to separate sciences. The spectroscope and the telescope with its modern adjuncts of photographic plates, photometer (which accurately measures the brightness of the light), bolometer (which measures the heat energy), and other physical instruments, have brought the methods of physics into astronomy, so that the great new science of astrophysics has arisen—new in the sense that it is practically the creation of the 20th century. It enables us to deduce the velocity of stars for which no motion can be seen with the telescope alone, and the rotation of the sun, planets, and nebulae, by modification of the spectral lines : to know of what elements the celestial bodies are composed, and at what temperature and in what electrical state these elements are : to arrange the stars in families and to discuss their evolution. These are achievements which, in the lifetime of men now living, would have seemed impossible. The external penetrating cosmic rays have been detected and investigated by methods developed in the physics laboratory. The latest application of the methods of physics in astronomy is the study of the radio waves reaching the earth from outer space, the new science of Radio Astronomy. The application of a variety of physical, mainly electrical, methods to chemical problems has given rise to the science of physical chemistry, likewise of comparatively recent date, and to-day successful attempts are being made to understand the simpler chemical combinations in terms of the latest physical discoveries about the properties of the atom, so that both the old organic and inorganic chemistry are tending to come increasingly within the scope of physics. The study of the way in which atomic nuclei are built up and break down has been called the New Chemistry. The geologist is more and more seeking explanations of his observed facts in terms of physical theory, so that the science of geophysics has recently developed. We may say that, even where the empirical facts of inanimate Nature are at present too complex for the application of precise physical theory, it is the ambition of those who are studying them to bring them ultimately within the scope of physics.

Law and Order. The object of physical science is to elaborate laws in terms of which the observed facts can be expressed—or explained, as long as we do not understand by this word any ultimate explanation—and new facts predicted. For, although it has been insisted that science is a collection of observed facts, facts without systemization on a theoretical basis do not constitute a science. This has been admirably expressed by the great mathematician Henri Poincaré, who, pointing out that we cannot be content with accumulating facts, says : " On fait la science avec des faits comme une maison avec des pierres ; mais une accumulation de faits n'est pas plus une science qu'un tas de pierres n'est une maison ",[1] and it might be added that a theory drawn up

---

[1] " We make science with facts, as we build a house with stones; but an accumulation of facts is no more a science than a heap of stones is a house."

without due regard for facts is no more a house than is an architect's draft. The point that a theory must be able to predict, Poincaré enforces by criticizing Carlyle's dictum that the fact is everything : that King John passed here is a reality such as alone matters, and for which, says Carlyle, I would give all the theories in the world. This is the language of the historian, retorts Poincaré ; the physicist should rather say : " King John has passed here : this is a matter of no interest to me, since he will not pass here again ".

Experiment and Explanation. Even a single experiment can never be carried out without a theoretical background from which criticism must be continually supplied. How otherwise can the experimenter decide what is significant and what is trivial ? His galvanometer gives an unexpected deflection. Is he to record it, to suppose that it was due to some chance disturbance, or to attribute it to something that he has overlooked in the plan of his experiment ? These are questions which he can often only solve by examining the event in the light of his theoretical knowledge of the origin of electrical currents. Or, even supposing that a clear-cut result is obtained, if it were not for the necessity of building up comprehensive theories, one experimental result would be as important as another, would be just another fact to be added to the store. It is the ability to conceive and carry out experiments which shall be theoretically significant, and to distinguish which facts, although facts, are petty and which fundamental, that characterizes the great experimenter. Men cannot be taught to be great observers any more than they can be taught to be great musicians. Rutherford, in his pioneer experiments on the scattering of alpha particles by thin foils, noticed that more particles were turned through very large angles than could be accounted for by the accepted view that the scattering was due to a large number of small random deflections. The proportion was very small, but it ought to have been even smaller. Rutherford might have neglected the few particles scattered through the large angles as due to some chance effect not worth investigating further, or he might merely have recorded it as a fact. Instead he saw that it could only be accounted for by some fundamental feature of atomic structure, concentrated attention on it, and from it derived the nuclear theory of the atom which dominates modern physics.

Often, too, the experimenter is guided by some inner conviction which resembles poetical inspiration or religious faith rather than the logical reasoning which is commonly supposed to be the main guide of the great originators in science. Faraday had an inner certainty that the fundamental physical manifestations were connected together, and, in particular, that light and magnetism were closely akin. In his own words : " I have long held an opinion almost amounting to conviction, in common I believe with many other lovers of natural knowledge, that the various forms under which the forces of matter are made manifest have one common origin ; or, in other words, are so directly related and mutually dependent that they are convertible, as it were, one into another, and possess equivalents of their power of action. . . . This strong persuasion extended to the powers of light, and led, on a former occasion, to many exertions, having for their object the discovery of the direct relation of light and electricity. . . ." It is thus that he commences his famous account of the experiments by which he established a connexion between magnetism and light, namely, the rotation of the plane of polarization of light passing through heavy glass in a magnetic field. For twenty-three years the subject of a relationship between magnetism and light occupied his mind before he finally achieved success. Faraday has been justly called the prince of experimenters, and it will be seen how large a part an inexplicable sense of what was to be expected played in his work. No doubt it has been equally strong in the case of other great men of science, but few have acknowledged it so plainly.

Scientific Method. A physical theory is a plan or scheme which accounts for certain observed facts in a way which gives us a quantitative connexion between them, and enables us to predict other effects which can be verified by experiment. It is particularly valuable in suggesting the lines along which further experiment can be carried out, so that a more satisfying knowledge of Nature can be won. If these experiments fail to give the anticipated results the theory has to be modified, until, as a rule, it is superseded by some more general theory. Quite often the newer work does not invalidate the older theory, but shows that its scope is somewhat less wide than was originally believed. This is, for instance, particularly the case with the theory of relativity, which is often supposed or suggested to have overthrown the Newtonian scheme of mechanics. Such is far from being the case. Newtonian mechanics maintains its unimpeached validity for material bodies in all ordinary circumstances, that is, except in the case of velocities which are only exceptionally observed for visible bodies, and then never on earth. Only in the case of one planet, Mercury, is the velocity sufficiently high to invalidate the Newtonian rules, and there the effect is only just within the limit of observation. Newtonian mechanics cannot be extended to such particles as swift electrons : its scope is not so universal as was believed some thirty years ago, but over a wide realm its reign is unbroken, and if it were not that the mechanics of relativity reduces to the Newtonian laws when ordinary velocities are in question it would never have gained acceptance.

The Unity of Nature. The whole tendency of science as it advances is to replace theories of limited validity by others more comprehensive, to reduce the number of general principles and extend their scope. This can be particularly seen in the way in which electrical theory has been elaborated and extended, so that to-day it tends to include all branches of physics. Faraday's conviction of the unity of Nature has been more and more justified. Clerk Maxwell, throwing Faraday's ideas into the mathematical form that lies at the basis of all modern theory of radiation, showed that waves of electromagnetic and magnetic force should travel with the velocity of light, and put forward the view that light was an electromagnetic wave. Hertz, under the influence of this theory, found out how to produce electromagnetic waves of the type of the wireless waves which are so familiar to-day. It soon became clear that all types of radiation which did not consist of particles could be united under one scheme, and grouped as electromagnetic waves whose different properties were due to differences of wavelength. When Röntgen discovered X-rays their nature was unknown : some twenty years after his discovery it was found that they too were electromagnetic waves, but of wavelengths in the neighbourhood of a few thousand-millionths of an inch, as against visible light of a few hundred-thousandths of an inch, and ordinary wireless waves of some hundreds of yards. One comprehensive electromagnetic spectrum includes all that aspect of Nature referred to as radiation, in contradistinction to matter.

Nature's Bricks and Mortar. On the other hand, the discovery of the electron, towards the end of the last century, permitted us to consider all ponderable matter as being made up of electric charges and Rutherford's scheme of the nuclear atom made this conception very much more precise. The theory that an atom of any given element consists of a central positively charged nucleus, containing practically the whole mass, surrounded by a cloud of electrons, enables the emission of light by matter to be worked out to a high degree of precision. This can only be done, however, by invoking an idea foreign to the physics of last century, an idea which is expressed in Planck's quantum theory of radiation. This theory states that radiation can be emitted only in discrete units of energy, a conception which is made particularly clear by Einstein's expression, " light darts ". Light has wave properties, but is sent out in individual spasms—" packets " or " darts ". This theory of Planck's has, of course, an experimental origin : it was put forward to account for peculiarities in the distribution of energy in the light sent out by a glowing solid. It seemed fantastic when it was first proposed, but it gave an explanation of this particular problem, which Planck saw was sufficiently important to warrant the drastic break with tradition which his scheme involved. Subsequent work has abundantly shown that the quantum theory has a universal validity in the minute world of atom dimensions, and it dominates modern theoretical physics.

Here we have something of profound significance for an understanding of the attitude of modern science. Physics up to the end of last century believed that all inanimate Nature could be explained by extending Newton's mechanics to the unseeably small—to atoms and ultimate particles, and to waves in an imaginary medium, the ether. This was the view of what is now called classical physics. Experiment has demonstrated that theories built on these lines will not work, and in consequence the more difficult task has been successfully attempted of working out the special rules which the ultimate particles and the whole range of waves obey. These rules cannot be illustrated with visible balls, rods, and strings, or with waves in water or air, for they are of an essentially different nature from those which govern tangible and visible particles. This fundamental fact constitutes one of the difficulties of explaining them. The situation can be fancifully illustrated by supposing a superhuman historian who could only see national forces at work—the migrations, wars, and decays of peoples. If he tried to apply the laws which he derived from these large-scale observations to individuals he might get a limited success in that both nations and single persons are subject to birth and death, but he would soon find it necessary to evolve a psychology of the individual which was quite different from that of the herd. Of course the herd is made up of individuals, so that ultimately the behaviour of the individual is all that matters, but in considering race movements it would be simpler for him to apply his mass laws than to try to work out the whole thing from the point of view of the single man. In the same way, when we know all about atoms we shall be able to express all bulk laws in terms of them, but in general it is simpler to stick to the particular rules which are valid for visible pieces of matter, that is, to classical mechanics and electro-dynamics, when we wish to know how, for example, machines and optical instruments behave.

The proton, or positively charged hydrogen nucleus, was until comparatively recently, 1932, supposed to be the only other kind of ultimate particle besides the electron, but research that year revealed a positive electron, which is the positive counterpart of the ordinary, or negative, electron ; and a neutron, or particle of the mass of the proton, but with no electric charge. The work of Aston on isotopes was particularly important as providing evidence that the masses of all the different atoms could be accounted for in terms of a nuclear structure of protons and neutrons. In 1935 further particles of mass intermediate between the proton and the electron were postulated by Yukawa to explain certain baffling properties of the atomic nucleus and were named mesons. The existence of mesons of different masses, charged and uncharged, and of very short life, has since been established experimentally. The exact relation between all these different particles is still obscure and involves wave conceptions to which reference is made in the next section, but we can say that we are confronted with a general scheme in which these fundamental particles and the forces between them play an essential part. Most of the researches in physical science and much of theoretical chemistry are to-day devoted to attempts to bring wide ranges of phenomena within this scheme.

It has been pointed out that waves have certain quantum or particle properties. It has further been shown by recent experiment that the ultimate particles, such as the electron, have certain periodic or wave properties. The antithesis between matter and radiation is, then, growing less marked, especially as, in Einstein's theory, which is generally accepted, radiation possesses mass and momentum. Much of atomic theory is to-day dominated by the theory known as wave mechanics, in which the interaction of fundamental particles and of radiation is explained in terms of vibrations of a kind expressed by certain equations which lie at the basis of the theory and constitute its authority. The structure of the nucleus, of the electronic part of the atom, the combination of atoms and the generation and absorption of radiation by atoms are all comprehended in this abstruse system. Gravitation, however, proves obstinate, for so far no attempt to group this force and the electric and magnetic force under one scheme, to devise one kind of field that shall comprehend all three, has proved very successful. The task of describing the gravitational force in terms of something more general baffled Newton and Faraday. Einstein has put forward a theory of a generalized field—gravitational, electric, and magnetic—but this theory does not suggest any possibility of an experimental test, which is the true touchstone of theories.

Science and Ethics. The latest advances of physics have not been without their philosophical repercussions. A close examination of the ultimates of physical theory, waves and particles, has shown that there must always be an element of indeterminacy in all attempts to fix either the position or the velocity of the particles. Some thinkers have sought to base upon this arguments as to the existence or non-existence of free will. The subject is very controversial, but it appears to the writer that scientific results can never have any bearing on moral problems of this nature. The scientific scheme is one devised particularly to deal with quantities subject to physical measurement—things that can be weighed, photographed, and recorded as numbers. It does not follow that it has any bearing on things that are not susceptible to numerical measurement and expression—in other words, the man of science abstracts the physically measurable from any problem, and deals with that. In any case Planck himself has clearly expressed the view that the concept of causality is something transcendental, and he admits free will because the individual ego is the one subject " where science and therefore every other causal method of research is inapplicable, not only on practical but also on logical grounds ". Einstein appears to be of the same opinion.

In the same way it may be contended that the time with which the theory of relativity deals is the astronomically measurable time, and that it does not follow that this is the only aspect of time, any more than the weight of an iron weight is the only aspect of that weight, although it may be the usual one. Dreams, for instance, suggest that time has other aspects than that measured by the clock. These are clearly thorny problems, but, whatever conclusion is ultimately reached, caution must certainly be used in any attempt to extend results obtained by laboratory methods to fields not contemplated in the elaboration of these methods. The tools were forged for working in one material : they may not be equally suitable for another material. An emotion, even if not measurable, has an existence as definite as a gram weight, but physics can deal only with the latter.

The Beauty and Utility of Mathematics. Mathematics stands in a peculiar position. It originated in the measurement and counting of material things, and a great deal of mathematical work is concerned with physical problems, with conceptions which arose in the laboratory, such as elastic, electric, and magnetic force. Nevertheless, some of the most fascinating and fundamental branches of mathematics deal with conceptions which have no counterpart in the physical world, and partake as much of the character of music as of science : the pattern and arrangement give a pleasure which apparently derives from some innate quality of the human intelligence. Geometries of space quite different from any measured space were worked out long before the theory of relativity (in which they seem to find an application) was thought of—of space, for instance, where the three angles of a triangle do not make up two right angles. They do not correspond to measured space, but they are self-consistent and intellectually satisfying. More abstruse examples exist, but their consideration would take us far afield. Just as a Persian carpet has two aspects, a beautiful pattern and a more pedestrian use, so mathematics has both an artistic and a scientific essence ; and just as some Persian carpets are hung on the wall, as too good for use as floor covering, so some branches of mathematics have been evolved without any reference to physical representation or use. Although so widely applied, mathematics has an aesthetic aspect and is to some extent so independent of the external world as hardly to be called a science.

BIOLOGICAL SCIENCES

The science of living matter has always been the subject of debate between two schools—the mechanists and vitalists. The old idea that the products of the organic world were of a nature totally different from those of the inorganic world, that there were material substances which life alone could produce, has long been shattered. Organic chemistry can make from dead matter large classes of colouring matter and flavours which at one time were only produced by the living plant, and the chemistry of the human body can be largely imitated in the test tube. The mechanism of nervous transmission, the chemistry of digestion, the physics of hearing—these are all things in which we have made advances and can certainly hope for more. The mechanical and chemical aspects of life can, then, certainly be explained and expressed in terms of physics and chemistry. This does not, however, mean that a man is merely an assemblage of mechanical and chemical systems in action. To understand the mechanism of a motor car is to know why it moves, but not why it gets to a particular place, and the vitalists point out that to account for animal behaviour and animal psychology is quite a different thing from working out a physical and chemical theory of nervous response. The vitalists, as represented by Professor McDougall, for instance, insist that the conception of purpose—teleology, as it is called—which is quite foreign to physical science, must be introduced into any fruitful discussion of animal behaviour, let alone human psychology : the mechanists hold that causal-analytical methods, the methods of the physical sciences, will eventually prove adequate. The subject is much too controversial to be resumed in a few words, but what is pretty certain is that the mechanists have not yet advanced very far towards their self-appointed goal.

SCIENCE AND HUMAN WELFARE

We have been discussing the province of pure science. There is, however, another aspect of enormous importance for our everyday lives, that of the applications of science. The growth of applied science is particularly clearly shown in the Universities, for whereas forty years or so ago it was taken for granted that practically all the best students in the scientific departments would go into one branch or another of academic life, to-day a large proportion is absorbed into the various industrial laboratories and associations which are engaged upon research for commercial firms or groups of firms. In fact, a new profession has grown up, and many students take their courses and proceed to training in research with industrial positions in view from the start. Some of the best brains of the country are now engaged in work of this kind, and, in addition to investigations of purely industrial interest, produce many results of wide scientific importance. These applications of physics and chemistry to our daily life are so many that it is impossible here even to glance at them : many of them will readily occur to any educated man. The applications of the biological sciences are no less important, for research in such subjects as soil science, the breeding of plants with required properties (such as disease-resisting wheats), and the breeding of cattle are revolutionizing agriculture. Very important is the study, initiated by Karl Pearson, of heredity by mathematical methods, for this can be applied to the human race, where deliberate breeding experiments are out of the question.

And the abuses of science ? Leaving out the invocation of the findings of science by narrow minds to support unworthy ends, one has but to say " War " to conjure up untold unhappiness that has been produced by engines wrought by science. This state of affairs, however, is less the fault of science than of an attitude of mind which can only be cured—how ? Certainly not by abolishing the work of science, which by dooming millions to starvation would at once produce world-wide conflicts. All things are subject to misuse : even the Holy Scriptures have been made an excuse for countless wars and for some of the worst horrors that have been recorded in history. Science and the scientific method have become not only a source of material prosperity but also an intellectual adventure which is a not unworthy feature of our age. Let us see to it that those who stand outside both the scientific and moral discipline do not direct the high pursuit of knowledge, which, apart from its loftier aspects, is fraught with untold material benefit for mankind, to baleful and pernicious ends.

E. N. DA C. ANDRADE

LEADING QUESTIONS

How and where did Science begin ?

Refer to pages 6 and 15.

What wonderful things was Galileo the first of all men to see ?

Pages 58 and 61.

What is the attitude of scientific men to " Seeing is Believing " ?

Page 27.

Are accepted scientific theories absolutely true ?

Page 33.

What secrets of the universe have been explained by recent discoveries in physics ?

Pages 72 and 81.

What gives Science its power in the modern world ?

Page 211.

Particles travelling at 17 miles a minute are continually hitting us. What are they ?

Page 145.

What is the electron ?   Matter or energy ?

Pages 190 and 202.

How can we make pictures of things far too small to be seen in the most powerful microscope ?

Page 289.

How did relativity modify Newton's system ?

Page 95.

Are there other worlds like ours ?

Page 108.

What mental rewards does the study of Science provide ?

Page 38.

BOOK I

THE ORIGIN
AND MEANING OF SCIENCE

WHAT IS SCIENCE?

SCIENCE is knowledge, but not all knowledge is worthy of the name of science. Before it can make this claim it must be organized and systematized. And, even more important, it must be tested and proved. *How far does the field of Science extend?* That is the first question we shall ask. The answer is given by Karl Pearson in the following lecture on "The Scope of Science". This extract is taken from *The Grammar of Science*, a book which Pearson built up round the critical lectures he gave at Gresham College, London, on the foundations of modern science.

KARL PEARSON

THE SCOPE OF SCIENCE

THE field of science is unlimited; its material is endless, every group of natural phenomena, every phase of social life, every stage of past or present development is material for science. *The unity of all science consists alone in its method, not in its material.* The man who classifies facts of any kind whatever, who sees their mutual relation and describes their sequences, is applying the scientific method and is a man of science. The facts may belong to the past history of mankind, to the social statistics of our great cities, to the atmosphere of the most distant stars, to the digestive organs of a worm, or to the life of a scarcely visible bacillus. It is not facts themselves which make science, but the method by which they are dealt with.

The material of science is co-extensive with the whole physical universe, not only that universe as it now exists, but with its past history and the past history of all life therein. When every fact, every present or past phenomenon of that universe, every phase of present or past life therein, has been examined, classified, and co-ordinated with the rest, then the mission of science will be completed. What is this but saying that the task of science can never end till man ceases to be, till history is no longer made, and development itself ceases ?

It might be supposed that science has made such strides in the last two centuries, and notably in the last fifty years, that we might look forward to a day when its work would be practically accomplished. At the beginning of the nineteenth century it was possible for an Alexander von Humboldt to take a survey of the entire domain of then extant science. Such a survey would be impossible for any scientist now, even if gifted with more than Humboldt's powers. Scarcely any specialist of to-day is really master of all the work which has been done in his own comparatively small field. Facts and their classification have been accumulating at such a rate, that nobody seems to have leisure to recognize the relations of sub-groups to the whole. It is as if individual workers in both Europe and America were bringing their stones to one great building and piling them on and cementing them together without regard to any general plan or to their individual neighbour's work ; only where someone has placed a great corner-stone is it regarded, and the building then rises on this firmer foundation more rapidly than at other points, till it reaches a height at which it is stopped for want of side support.

Yet this great structure, the proportions of which are beyond the ken of any individual man, possesses a symmetry and unity of its own, notwithstanding its haphazard mode of construction. This symmetry and unity lie in scientific method. The smallest group of facts, if properly classified and logically dealt with, will form a stone which has its proper place in the great building of knowledge, wholly independent of the individual workman who has shaped it. Even when two men work unwittingly at the same stone they will but modify and correct each other's angles. In the face of all this enormous progress of modern science, when in all civilized lands men are applying the scientific method to natural, historical, and mental facts, we have yet to admit that the goal of science is, and must be, infinitely distant.

For we must note that when from a sufficient if partial classification of facts a simple principle has been discovered which describes the relationship and sequences of any group, then this principle or law itself generally leads to the discovery of a still wider range of hitherto unregarded phenomena in the same or associated fields. Every great advance of science opens our eyes to facts which we had failed before to observe, and makes new demands on our powers of interpretation.

This extension of the material of science into regions where our great-grandfathers could see nothing at all, or where they would have declared human knowledge impossible, is one of the most remarkable features of modern progress. Where they interpreted the motion of the planets of our own system, we discuss the chemical constitution of stars, many of which did not exist for them, for their telescopes could not reach them. Where they discovered the circulation of the blood, we see the physical conflict of living poisons within the blood, whose battles would have been absurdities for them. Where they found void and probably demonstrated to their own satisfaction that there was void, we conceive great systems in rapid motion capable of carrying energy through brick walls as light passes through glass. Great as the advance of scientific knowledge has been, it has not been greater than the growth of the material to be dealt with.

The goal of science is clear—it is nothing short of the complete interpretation of the universe. But the goal is an ideal one—it marks the *direction* in which we move and strive, but never a stage we shall actually reach. The universe grows ever larger as we learn to understand more of our own corner of it.

It seems that Science will have no end—at least while mankind exists on the earth. What about its beginning? *How and where did Science begin?* Remembering that science is organized knowledge it is easy to understand that it could only originate when man had sufficient leisure, after the satisfaction of his physical wants, to ponder over the problems placed before him by his observations of the world around. How science was born of man's need to know and man's need to do is explained in the musical prose of the following extract from a lecture by J. L. Myres. This formed part of a summer school Course on " Science and Civilization ", the sixth in a series on the Unity of Civilization.

SIR JOHN L. MYRES

THE BEGINNINGS OF SCIENCE

CIVILIZED people's " thinking ", as distinct from mere use of their brains in the daily round of their lives, is concerned mainly with high matters which they do not quite understand yet, but their ignorance of which is beginning to be felt practically as an obstacle to desirable activity. Thinkers among ourselves think, for example, about the constitution of matter; the nature of force and life; high problems of conduct, of politics, of mental and metaphysical philosophy.

But the thinking of the savage, too, is concerned with high matters, too high as yet for him, but of no less practical bearing : with hunger, sickness, rain; with the baffling ways of game, and the behaviour of dogs and womankind; with his own feelings and imaginations; and among them, above all, with his dreams. All these puzzle him and prevent him from living as well as he has it in him to live under his actual conditions. He needs, that is, and apparently strives to achieve, in his thinking, two things mainly; an explanation, intelligible to himself, of what it is that is going on; and direction, how to behave, in presence of this which is going on in Nature and in his own experience as a whole.

The explanation is to be a response to impulses of curiosity—that is, to his *need to know*—as vital to his well-being as those of any young animal, or of a modern baby, or researcher ; curiosity, most active in modern man during infancy and adolescence, and customarily repressed in most of us ; but lifelong whenever it has not been so repressed, and irrepressible, as we know, in idiots, gossips, and men of science.

From the need of explanation, that is, comes all pure science, no less than all true poetry. From the need of direction, on the other hand, all applied sciences : For the need of direction stands in similar relation to man's *need-to-do*, his instinct to push, to hustle, to disarrange and rearrange things about him till he has got them arranged as he likes, from which all material arts and technical skill proceed, for the conquest and domestication of intractable Nature ; and, no less, all social arts and political skill, for the conquest and domestication—let us say at once the " civilization "—of no less intractable man. Such thought with a view to action involves foresight, and leads to invention ; as thought with a view to knowledge presumes imagination, and achieves discovery.

In such thought with a view to knowledge, there are two distinct phases : in the first, the mere facts are established by observation, and up to the point where their own natural needs are concerned, quite primitive men are good observers, and may also be good recorders of what they observe. The Bushman drawings and the wonderful naturalism of the old French and Spanish cave-dwellers are ample evidence for this. But mere facts have never taken thinking men very far on the road to knowledge ; and the second phase of thought begins when the meaning of the fact is expressed by what the Greeks knew as *hypothesis*, by *supposition* that is, or in plain English *underpinning* ; filling in, beneath and behind the facts, whatever is conceived in imagination as really going on, and presenting these appearances to the observer. Primitive ingenuous man, like the greatest philosophers, is but trying to get beneath the surface, behind the veil of appearance, and to reach reality. . . .

[His attempts, however, especially in areas where life is hard and difficult, are hampered by Fear and by Pride, and to these are due his failure to push beyond hypothesis to explanation.]

A savage, resting in his cave on a windy night, hears howling outside. Nothing is to be seen to account for the noise, but it—whatever *it* is—howls like a wolf : or rather (he thinks) it is not an ordinary wolf ; but if a wolf were big and fierce enough, it would howl like that. . . .

Our savage knows quite enough about wolves to avoid going out at night while howling goes on. With such a howling, common prudence argues irresistibly against experimental verification : and it is at this point, of omission to test and (in the older sense) " prove " the hypothesis, that mythology has its frontier with science : a myth being unverified hypothesis of what is really happening, as well as symbolic notation for what appears to happen, such as poets and men of science alike employ.

And the paralysis, wrought in this instance by fear, may be wrought no less by pride, in the not unlikely event of the wind-wolf notion working in the mind of its originator, in mere consolation of his fear, or in emotional enhancement of his apprehension of the noise as wolf-noise, so that conviction seizes him, or perhaps seizes first his fellows in the cave, and the mere achievement of such a guess, like any other creative activity, satisfies his sense of need. . . .

In contrast, we may look to those rare regions where the balance sets so far in favour of Man as to allow not merely those adaptations of animals and plants to his uses which are common to tribally organized societies throughout the north temperate zone and far into the less amenable districts on either margin of it, but even what we may describe as the first great domestication of inanimate nature, the redistribution of water over the land surface in accordance with man's will, to make plants grow and animals breed in a land which is his by right of creation. Such domestication of water is the fundamental achievement on which are based the great river-valley cultures and the vast aggregations of humanity, which these quite artificial conditions rendered possible.

So elaborate and extensive an organization as a riparian state required, and achieved, accurate means to communicate instructions and register events : it is the same *need to do* and *need to know* as we have seen already giving rise on a lower social plane to myth and magic. Longer foresight demands accurate measurement of time and a calendar-scheme for record and for programme. Wider conceptions of area and cubic content, whether for earthen-dams, or water-flow, or estimation of produce and tribute, lead to systems of measurement, and eventually even of weight and value. To this extent the mere needs of administration enforced and stimulated new notions of quantitive accuracy, of permanent and world-wide significance. No less important is the reaction of the political and economic régime on the cosmology, and the connexions between " physics and politics " deserve closer examination.

The peculiarities of the Egyptian interpretation of nature may be illustrated by analysis of some of its leading ideas. In the Nile valley, man's struggle against the wild succeeds on two conditions, of industry and observation. Physical contrasts of seasonal and regional fertility are abrupt ; solar heat contends with Nile water, sea breeze with scorching *hamseen*. Man must discern which of the Powers are his friends, and serve them ; and in such a world-war, even the Greater Powers recognize service rendered. First of all Powers, as already hinted, are Sun and River ; then the Bull, Cow, and Ram which his leaders brought with them, or reclaimed from the wild. These are beneficent, but require observation and maintenance, so near the margin of pasturage. Others are maleficent, and need observation too ; the crocodile, scorpion, uræus-snake. Others challenge wonder, by incessant or periodic or exceptional activity ; the hawk flying in the eye of the sun, the dung-beetle rolling his pellet—whither ? and why ?—the jackal, guarding (or is he molesting ?) the dead ; the ibis, ubiquitous and all-seeing ; the cat, " for all places are alike to him " ; lion, hippopotamus, and many more ; all elements of a ramshackle universe, precariously held together, like the baronies of Delta and Valley by the wearer of the double crown, who is " child of the Sun ", and co-regent over men with him, as his own son is his co-regent.

Large matters are " explained " by the simplest of diagrams ; creation, in a world of Nile-mud, is potter's business at bottom ; the sun's daily course is a boat journey through the sky, blue as the river's reflection of it. Complex activities are surely the work of composite deities. If the same functions are performed in Crocodilopolis by King Crocodile and in Bubastis by Our Lady Puss, reason points to that which in cat and crocodile is one, as the generic cause. And all these powers, working as reasonably as they do, are in some sense human ; it is not mere ibis or lion, but an intelligent and intelligible energy like man's ; and the animal head surmounts a body of which the working mechanism is human. Above all, and gradually subsuming and harmonizing this polytheism, two master-conceptions dominate and at times compete : a naturalist hypothesis, attributing all ultimately to solar-energy ; and a humanist hypothesis, that all growth and maintenance is ultimately congruous with the birth and life and death of a man, the only sequence in all nature, as we have seen in the most primitive phase, of which man directly knows the inwardness.

And so Egyptian knowledge culminates in the facts of motherhood and childhood—Isis and Osiris—ever bearing, ever-born, and ever born again, as day succeeds night, and awakening sleep, and life death, when that which is not-us wins periodic victory over laborious beneficent manhood. With this Osiris-mystery, as with Egyptian civilization itself, we pass beyond indigenous Nile-dom, for Osiris is shepherd-king ; in another of his aspects he is lord of some un-Egyptian

tree, which is his written symbol; in another, his energy
is vegetative rather than animal, and in an essentially agricultural régime analogies between the fate of seed-corn and that
of man himself, desiccated and buried away " till Osiris shall
come ", acquired profound significance, and suggested infinite
precaution to conserve that unpromising remnant of *my*
career, in mummy-shroud and sarcophagus, pyramid-tomb
and chantry priest, against that far renaissance.

Thus Egyptian attempts to rationalize the Nile-world led
first to ill-harmonized analyses of nature-forces; then to
loose synthesis round twin hypotheses of solar and vital
energy—Ra and Osiris—nowhere carried really farther than
political theory was carried by the ramshackle Egyptian
administration and social structure; and then the lurking
remnant of primæval fear—fear of my own dissolution in
spite of all—imposed the dead hand of vast insurance-societies,
the great priestly guilds, on further progress in administration,
in society, in economics and industry. For with a cosmology
so complicated and so incomplete, only an expert could understand the symbolism, much less perform the ritual.

As elsewhere, too, repeated failures of the experts themselves led to the conviction that human insufficiency through
latent defect in the client—not of course in the procedure or
its theory!—must be almost irremediable: so that both
ingenuity and wealth were diverted from more accurate
interrogation of nature into refinements of current ceremonies.
Thus the " wisdom of the Egyptians " forwent enhancement
of this life, and spent itself to ensure a sequel. It is perhaps
not without significance that those later systems of knowledge
which have been imbued most deeply with that wisdom have
been most liable to the grip of the same " dead hand ".
Lucretius feared this when Isis had but newly come to Rome:
Piers Plowman heralds more effective " protest " against it.

In its outlook on nature the other great river-culture of the
Ancient East presents curious similarities with that of Egypt,
and also instructive contrasts. In an artificial country like

Babylonia, terrestrial nature is poor : " in the beginning no
reed was ; no tree ; no dry land for cities " : only mudflats
like those which infest the Euphrates delta to-day. Only the
sky-phenomena are copious in quantity (the sky being usually
clear), though simple in kind, being essentially movements of
points. Once correlated empirically (which was fairly easy)
with the cycle, less accurately graded, of water-flow and
weather and especially of the seasonal winds, the movements
of the stars were seen to be primary, and all else coherent with
them. The " great powers ", Sun, Moon, Planets, and Stars,
have however a less and less orderly escort (like the courtiers
and camp-followers of a king) of Winds and Storms " fulfilling
his word ", Waters and Sandbanks, the latter rebellious,
incalculable, " disastrous " in the literal sense that there was
no *star* that really looked after them. Then there are domestic
animals, with biddable natures like those of men, and plants
whose innate observance of the kindly seasons brought them
under the sweet influence of the Pleiades or whatever constella-
tion brought up " herb for the use of men " upon the earth ;
and there were wild unbiddable blights and vermin, serpents,
scorpions, and poison-spiders most pernicious of all. These
too, like the hangers-on of temporal majesty, were to be
observed, selfishly, with precaution ; and bound over to keep
the peace, by a " word of might " from one of the greater gods.

Now a " great god ", like a great priest-king in a temple-
city, was a public benefactor, an earthly providence. At his
will he could interrupt the order of your days, cut off your
water, commandeer your seed-corn ; or turn again and bless
you with freehold land, and a place at his table. But if he was
an earthly providence, he was before all things a capitalist :
the earth was his, for he made it, out of mud and primæval
slime ; the water was his, to irrigate the just and the unjust ;
wealth was his to give or to withhold. But all things had their
price in Babylonia ; the talent, lent out of the temple treasury,
to be returned at the day of reckoning with interest as per
agreement, was counterpart and symbol of other " talents "

of character and skill, to be employed as under the master's eye.

Thus in Babylonia, and rather less clearly and crudely in Egypt, much that at an earlier phase was paraphrased by myth is now found amenable to explanation, capable of being verified by recurring experience ; and this is mainly due to the circumstance that in organized society, with adequate means of *record*, observers multiply in each generation, and observations can accumulate in time. Natural history, at all events, is found to repeat itself ; a general order, intelligible to man, is made out, and attributed to the foresight and administration of the Great Powers. Thence emerges a hypothesis of Providence, and a formulation of the Will of the Gods, which for man are laws, like the laws of Hammurabi : we have echoes of such an astronomical Digest in the Hundred-and-fourth Psalm, " He hath appointed the moon for certain seasons, and ` the sun knoweth his going down " ; more distant echo still in the fragment of Heraclitus, " the sun will not overstep his landmarks ; if he do so, avenging spirits, minions of the law shall find him out ". But the " Code " is not absolute, nor complete ; the gods of a Theocracy do not yet find all their work " very good " ; any more than Khammurabi or Ur-kagina may rest from their labours. The " great dragon underground ", or rather in the lower fens, is not yet dead : his head is bruised but not broken by the " seed of the woman ". Only by divine foresight—let us call it by its Roman name of " Providence " while noting that it is neither all-seeing, nor all-powerful yet—can man preserve from bruises his own Achilles' heel.

Providence, that is, is reinforced with Miracle ; for as long as miracle is possible, Equity supplements the Code ; Justice is tempered with Mercy. And as long as Mercy, Equity, Miracle, remain, Magic remains too ; the possibility, namely, of somehow getting the judge on to the right side, or yourself on to the blind side of him. Thus with an organized religion, we find magic organized too, and the name *mage* and *magic*, like the Roman equivalent *Chaldæan* and our own *Gipsy* (that is, *Egyptian*), mark the modes in which this organized Magic penetrated into the West.

Science did not flourish in its cradle in the bulrushes. There it remained the monopoly of the priests, and the atmosphere was not suitable for its growth. But farther north in the isles of Greece the winds of curiosity and inquiry blew more freely, and it was there that facts were first accumulated and subjected to man's reasoning faculties, so that they gave something bigger and more valuable than their mere sum.

In the earlier half of the sixth century B.C. a man called Thales flourished (financially as well as in other ways) in Miletus, a Greek city on the west coast of Asia Minor. As a business man he visited Mesopotamia and Egypt. From Babylon he brought back sufficient astronomical knowledge to enable him to predict an eclipse of the sun on May 28, 585 B.C. In Egypt he picked up sufficient practical knowledge of triangles to enable him to lay the foundations of geometry by giving general rules instead of the special cases which had sufficed for the Egyptian surveyors. Some of the earliest theorems which we learnt at school were first proved by Thales : *e.g.*, that when two lines intersect one another the opposite angles are equal ; or that the angles at the base of an isosceles triangle are equal. However, it was not the fact that he proved this property of the isosceles triangle, which was important, so much as the way in which he did it. He could not prove it by just looking at the triangle. He had to make an experiment ; he had to draw a " construction ", making a line bisect the angle at the apex, and producing this until it met the base, thus dividing the original triangle into triangles which could be proved equal. From this it followed by common-sense arguments, and without the assumption of any new properties, that in *every* triangle which has two equal sides, the angles opposite those sides must be equal. *Why may we regard this as the first step in pure science ?* Its importance will be brought home to us by the following extract, freely translated, from Kant's preface to the second edition of his *Critique of Pure Reason*.

IMMANUEL KANT

THE FIRST STEP IN PURE SCIENCE

IN the earliest days of which history has given us a record, that is to say with that wonderful people the Greeks, Mathematics had already started to travel along the sure road of a *science*. It must not be supposed that it was as easy for mathematics to find, or rather to build, this royal road as it was for logic, where reason is concerned with itself alone. On the contrary I believe that with mathematics it long remained a case of blind groping about in search of its true aims and destination. The Egyptians in particular were still in that stage. The transformation was the result of the happy idea of one man who was inspired to try an experiment; and from this point onward the road lay straight ahead, inevitable and endless.

The exact history of this intellectual revolution—much more important in its results than the discovery of the passage round the Cape of Good Hope—has not been preserved, nor do we know much about its author. But Diogenes Laertius, in naming the supposed discoverer of the way to prove some of the simplest of geometrical theorems—theorems which hardly seem to need any proof to us now—makes it very clear that the vistas opened up by a new technique of geometrical construction and proof were regarded as of the most vital importance by the mathematicians of that bygone age. A new light must have flashed on the mind of the man (Thales or whatever his name may have been) who first found a way of demonstrating the essential property of an isosceles triangle. For he found that it was not sufficient to meditate on the figure as it lay before his eyes, or to brood over a mental image of it. It was not thus that the property of the triangle could be made clear. He tried an experiment; he made a *construction* and it was the construction which brought the property of the isosceles triangle within range of his arguments and his previous knowledge of lines and angles. It was this inspired idea which opened up the endless vistas.

A much longer period elapsed before Physics entered on the highway of Science—when the wise Bacon gave a new direction to physical studies, or rather—as others were already on the right track—imparted fresh vigour to the pursuit of this new direction. Here too, as in the case of mathematics, there is evidence of a rapid intellectual revolution. . . .

When Galileo experimented with balls of different known weights on an inclined plane, when Torricelli caused the pressure of the air to sustain a weight which he had calculated beforehand to be equal to that of a definite column of water, or when Stahl, at a later period, converted metal into lime, and lime into metal by the addition and subtraction of certain elements, a light broke upon all natural philosophers. They learned that reason only sees what it is looking for. It must not be content to follow humbly in the leading-strings of nature ; it must push on ahead with principles of judgment according to unvarying laws, and it must compel nature to reply to the questions which it puts. For accidental observations, made according to no preconceived plan, cannot be united under a universal law. Yet it is this necessary law which reason seeks, and which it must find. Reason deduces the natural law from the observations which suggest and support it, and it is only when experiment is directed by reason that it can serve any useful purpose.

Reason, indeed, must approach nature, not in the guise of a pupil who listens to all his master chooses to tell him, but rather in that of a judge, who compels the witnesses to reply to those questions which he himself thinks fit to propose. This is the inspiring idea through which, after groping in the dark for so many centuries, natural science was at length led into the path of solid and continuous progress.

How exciting the quest must have been in those early days of Science ! The whole of Nature was spread out before the questioning eyes of the Greeks, all asking to be explained. Thales had a multitude of successors : mathematicians, astronomers, geographers, biologists, and medical men, although we give a rather false impression by classifying them thus. The Greek scientist was not a specialist ; he was always seeking for universal explanations which would throw light on our knowledge of the world as a whole. He studied fishes or planets or numbers, not only for themselves, but, still more important, for what they might tell him about the whole of " nature " and the universe. How did they make their discoveries ? *How does Science seek Truth ?* The answer, in another extract from Karl Pearson's *Grammar of Science,* may surprise you, but it is none the less true. It is the artist in the scientist who makes the great discoveries ; it is the man of imagination who sees the hidden law behind the mass of facts.

KARL PEARSON

SCIENCE AND THE IMAGINATION

THERE is another aspect from which it is right that we should regard pure science—one that makes no appeal to its utility in practical life, but touches a side of our nature which the reader may have thought that I have entirely neglected. There is an element in our being which is not satisfied by the formal processes of reasoning ; it is the imaginative or æsthetic side, the side to which the poets and philosophers appeal, and one which science cannot, to be scientific, disregard.

We have seen that the imagination must not replace the reason in the deduction of relation and law from classified facts. But, none the less, disciplined imagination has been at the bottom of all great scientific discoveries. All great scientists have, in a certain sense, been great artists ; the man with no imagination may collect facts, but he cannot make great discoveries. If I were compelled to name the Englishmen who during our generation have had the widest imaginations and exercised them most beneficially, I think I should put the novelists and poets on one side and say Michael Faraday and Charles Darwin.

Now it is very needful to understand the exact part imagination plays in pure science. We can, perhaps, best achieve this result by considering the following proposition : Pure science has a further strong claim upon us on account of the exercise it gives to the imaginative faculties and the gratification it provides for the æsthetic judgment. The exact meaning of the terms " scientific fact " and " scientific law " will be considered in later chapters, but for the present let us suppose an elaborate classification of such facts has been made, and their relationships and sequences carefully traced. What is the next stage in the process of scientific investigation ? Undoubtedly it is the use of the imagination. The discovery of some single statement, some brief *formula* from which the whole group of facts is seen to flow, is the work, not of the mere catalogue, but of the man endowed with creative imagination. The single statement, the brief formula, the few words of which replace in our minds a wide range of relationships between phenomena, is what we term a scientific *law*. Such a law, relieving our memory from the burden of individual sequences, enables us, with the minimum of intellectual fatigue, to grasp a vast complexity of natural or social phenomena.

The discovery of law is therefore the peculiar function of the creative imagination. But this imagination has to be a *disciplined* one. It has, in the first place, to appreciate the whole range of facts which require to be resumed in a single statement ; and then, when the law is reached—often by what seems solely the inspired imagination of genius—it must be tested and criticized by its discoverer in every conceivable way, till he is certain that the imagination has not played him false, and that his law is in real agreement with the whole group of phenomena which it resumes.

Herein lies the key-note to the scientific use of the imagination. Hundreds of men have allowed their imagination to solve the universe, but the men who have contributed to our real understanding of natural phenomena have been those who were unstinting in their application of criticism to the product of their imaginations. It is such criticism which is the essence of the scientific use of the imagination ; which is, indeed, the very life-blood of science.

No less an authority than Faraday writes :—

" The world little knows how many of the thoughts and theories which have passed through the mind of a scientific investigator have been crushed in silence and secrecy by his own severe criticism and adverse examination ; that in the most successful instances not a tenth of the suggestions, the hopes, the wishes, the preliminary conclusions have been realized."

The reader must not think that I am painting any ideal or purely theoretical method of scientific discovery. He will find the process described above accurately depicted by Darwin himself in the account he gives us of his discovery of the law of natural selection. After his return to England in 1837, he tells us, it appeared to him that :—

" By collecting all facts which bore in any way on the variation of animals and plants under domestication and nature, some light might perhaps be thrown on the whole subject. My first note-book was opened in July 1837. I worked on true Baconian principles, and, without any theory, collected facts on a wholesale scale, more especially with respect to domesticated productions, by printed inquiries, by conversation with skilful breeders and gardeners, and by extensive reading. When I see the list of books of all kinds which I read and abstracted, including whole series of Journals and Transactions, I am surprised at my own industry. I soon perceived that selection was the keystone of man's success in making useful races of animals and plants. But how selection could be applied to organisms living in a state of nature remained for some time a mystery to me."

Here we have Darwin's scientific classification of facts, what he himself terms his " systematic inquiry ". Upon the basis of this systematic inquiry comes the search for a law. This is the work of the imagination ; the inspiration in Darwin's case being apparently due to a perusal of Malthus's *Essay on Population*. But Darwin's imagination was of the disciplined, scientific sort. Like Turgot, he knew that if the first thing is to invent a system, then the second is to be disgusted with it. Accordingly there followed the period of self-criticism, which lasted four or five years, and it was no less than *nineteen* years before he gave the world his discovery in its final form. Speaking of his inspiration that natural selection was the key to the mystery of the origin of species, he says :—

" Here, then, I had at last got a theory by which to work ; but I was so anxious to avoid prejudice, that I determined not for some time to write even the briefest sketch of it. In June 1842 (*i.e.* four years after the inspiration), I first allowed myself the satisfaction of writing a very brief abstract of my theory in pencil in 35 pages ; and this was enlarged during the summer of 1844 into one of 230 pages, which I had fairly copied out and still possess."

Finally an abstract from Darwin's manuscript was published with Wallace's Essay in 1858, and the *Origin of Species* appeared in 1859.

In like manner, Newton's imagination was only paralleled by that power of self-criticism which led him to lay aside a demonstration touching the gravitation of the moon for nearly eighteen years, until he had supplied a missing link in his reasoning. But our details of Newton's life and discoveries are too meagre for us to see his method as closely as we can see Darwin's, and the account I have given of the latter is amply sufficient to show the actual application of scientific method, and the real part played in science by the disciplined use of the imagination.

We are justified, I think, in concluding that science does

not cripple the imagination, but rather tends to exercise and discipline its functions. We have still, however, to consider another phase of the relationship of the imaginative faculty of pure science. When we see a great work of the creative imagination, a striking picture or a powerful drama, what is the essence of the fascination it exercises over us ? Why does our æsthetic judgment pronounce it a true work of art ? Is it not because we find concentrated into a brief statement, into a simple formula or a few symbols, a wide range of human emotions and feelings ? Is it not because the poet or the artist has expressed for us in his representation the true relationship between a variety of emotions, which we, in a long course of experience, have been consciously or unconsciously classifying ? Does not the beauty of the artist's work lie for us in the accuracy with which his symbols resume innumerable facts of our past emotional experience ? The æsthetic judgment pronounces for or against the interpretation of the creative imagination according as that interpretation embodies or contradicts the phenomena of life, which we ourselves have observed. It is only satisfied when the artist's formula contradicts none of the emotional phenomena which it is intended to resume.

If this account of the æsthetic judgment be at all a true one, the reader will have remarked how exactly parallel it is to the scientific judgment. But there is really more than mere parallelism between the two. The laws of science are, as we have seen, products of the creative imagination. They are the mental interpretations—the formulæ under which we resume wide ranges of phenomena, the results of observation on the part of ourselves or of our fellow-men. The scientific interpretations of phenomena, the scientific account of the universe, is therefore the only one which can permanently satisfy the æsthetic judgment, for it is the only one which can never be entirely contradicted by our observation and experience. It is necessary to emphasize strongly this side of science, for we are frequently told that the growth of science is destroying

the beauty and poetry of life. It is undoubtedly rendering
many of the old interpretations of life meaningless, because it
demonstrates that they are false to the facts which they profess
to describe. It does not follow from this, however, that the
æsthetic and scientific judgments are opposed; the fact is,
that with the growth of our scientific knowledge the basis of
the æsthetic judgment is changing and must change. There is
more real beauty in what science has to tell us of the chemistry
of a distant star, or in the life-history of a protozoon, than in
any cosmogony produced by the creative imagination of a pre-
scientific age. By " more real beauty " we are to understand
that the æsthetic judgment will find more satisfaction, more
permanent delight, in the former than in the latter. It is this
continual gratification of the æsthetic judgment which is one
of the chief delights of the pursuit of pure science.

There is an insatiable desire in the human breast to resume
in some short formula, some brief statement, the facts of
human experience. It leads the savage to " account " for all
natural phenomena by deifying the wind and the stream and
the tree. It leads civilized man, on the other hand, to express
his emotional experience in works of art, and his physical and
mental experience in the formulæ or so-called laws of science.
Both works of art and laws of science are the product of the
creative imagination, both afford material for the gratification
of the æsthetic judgment. It may seem at first sight strange
to the reader that the laws of science should thus be associated
with the creative imagination in man rather than with the
physical world outside him. But, as we shall see, the laws of
science are products of the human mind rather than factors of
the external world. Science endeavours to provide a mental
résumé of the universe, and its last great claim to our support
is the capacity it has for satisfying our cravings for a brief
description of the history of the world.

Such a brief description, a formula resuming all things,
science has not yet found and may probably never find, but
of this we may feel sure, that its method of seeking for one is

the sole possible method, and that the truth it has reached is the only form of truth which can permanently satisfy the æsthetic judgment. For the present, then, it is better to be content with the fraction of a right solution than to beguile ourselves with the whole of a wrong solution. The former is at least a step towards the truth, and shows us the direction in which other steps may be taken. The latter cannot be in entire accordance with our past or future experience, and will therefore ultimately fail to satisfy the æsthetic judgment. Step by step that judgment, restless under the growth of positive knowledge, has discarded creed after creed, and philosophic system after philosophic system. Surely we might now be content to learn from the pages of history that only little by little, slowly line upon line, man, by the aid of organized observation and careful reasoning, can hope to reach knowledge of the truth, that science, in the broadest sense of the word, is the sole gateway to a knowledge which can harmonize with our past as well as with our possible future experience. As Clifford puts it, " Scientific thought is not an accompaniment or condition of human progress, but human progress itself ".

The Greeks had not Darwin's patience. Science was young and the world was there and they were eager to explain it. Nor had they Faraday's experience of the complexities of nature, and of the need for caution in accepting theories which seemed more or less to fit the facts. They were optimistic and they were confident. So it was easy for them to make mistakes, and of course they did make many. They were not alone in that, and when we remember that they had to start from scratch, their occasional stumbles fade into insignificance compared with the real progress which they made. The greatest name among the Greek scientists is that of Aristotle, and his main contribution was in the field of biology. But even in the physical sciences, where he often went astray, he did lay down the " royal road " which had to be followed if enduring progress was to be made. *How did Aristotle blaze the trail for scientific progress ?* A few brief extracts from his *Physics* will make this clear and give us some idea of his mettle.

ARISTOTLE

THE PRINCIPLES OF NATURAL SCIENCE

K NOWLEDGE or understanding comes only when we have discovered first principles, that is the ultimate causes which produce the effects, and the elements of which things are made. In the study of nature we must strive to discover and establish principles which all things obey.

The path of investigation is clear ; we start with what is easily observed even if it is not so easily understood. From these observations we must work back to the principle which will illuminate them for us and so give us the deeper knowledge we seek. It is the concrete and particular which is easily observed and grasped ; the abstract and general must be deduced from our earlier observations and our analysis of them. . . .

Some things exist, or come into existence, by nature ; some in a different way. Animals, plants, and the elements— earth, fire, air, and water—these, and things like them, exist by nature. And they are different from things which are not so constituted because they all seem to have in them a principle of movement, or change, and rest. The change is sometimes in size, as in growth or shrinkage, and sometimes in quality or appearance. Non-natural things, however, such as a bedstead or a garment, have no such inherent tendency towards change, except in so far as they are made of natural things, and partake of the tendency of such things to change. . . . If a man buried a bedstead and it germinated and threw up a shoot, it would be a tree and not a bedstead that would come up. Again, men propagate men, but bedsteads do not propagate bedsteads ; and so we say that the natural factor in a bedstead is not its shape—that is art—but the wood. The artificial shape

is art ; the matter which does reproduce itself—man or wood—
is nature. "Nature" is etymologically equivalent to "genesis".
" Nature " comes from the same root as " nativity ". . . .

We often blame fortune or luck or accident as the cause of
events. Is there any difference between fortune and accident ?
And in any case what are they ? Some people say there is no
such thing as luck or accident ; they assert that everything has
its cause, and that whenever we blame chance there is always
some other cause to be found. And indeed it does seem that
if there be such a thing as luck it should be capable of study
and elucidation by the philosophers. It is, however, note-
worthy that even the men who deny the existence of chance
happenings in theory are always ready, in practice, to distinguish between things that do and those that do not depend
upon chance or luck.

Some indeed say that the firmament and all the worlds
came by chance, through the whirling and shifting that
separated the cosmos from the chaos. How strange, that mean
little things like animals and plants should have a definite
cause for their existence while the heavens and the divinest
things that can be seen by man should come just anyhow !
It is just in the movements of these heavenly bodies that we
never see casual or accidental variations. Fortune, however,
may be a genuine cause of things, but one that has something
divine and mysterious about it so that it is beyond human
comprehension. . . .

Another subject for study is the Infinite or the Unlimited.
We believe in this for the following reasons : (1) Time seems
to be unlimited. (2) Mathematicians treat magnitudes as
infinitely divisible. (3) There seems to be an unfailing store-
house of life. (4) Since things which are limited are only so
limited by coming up against something else, how can there be
any absolute limit ? (5) And, most important of all, the
imagination can always conceive a " beyond " reaching out
further than any limit, just as there is always a number bigger
than any number one cares to name.

This, however, is a genuine problem, for whether we admit or deny the existence of the Unlimited we find ourselves in an untenable position. A " number ", for example, is something that can be counted, so it cannot be infinite. And if you have an infinite body how can it move ? Where can it move to ? How can there be an " up " and a " down " or a " left " and a " right " in infinite space ? For these distinctions are not merely relative to the observer, but are defined in the universe itself. For " down ", wherever one may be, means in a straight line from thence to the centre of the earth, while " up " means in a straight line in the other direction.

As for Time, it owes its continuity to the " now ", and yet this same " now " divides it into present and past. Time is the dimension proper to movement, and it is " now " that moves along it.

The Greeks often found themselves in an " untenable position " in their attempts to answer all the urgent questions which crowded their inquiring minds. And at quite an early stage they began to feel the need for some touchstone of truth, some unfailing acid test which would separate the pure gold of truth from the dross of deception. They did not find it, although Socrates gave them an alternative—the avoiding of self-contradictions—which afforded a workable means of progress. Their first instinct, of course, was to rely on their five senses, but they soon learnt to distrust these. *What is the attitude of scientific men to the maxim " Seeing is Believing " ?* As Sir Richard Gregory said, this may be sound enough doctrine for the majority of people, but it is insufficient as a principle of scientific inquiry—as will be very clearly illustrated later on in this volume. Even a round table only looks round when viewed from a point directly above or below its centre. So scientists adopt a policy of prudent caution, even though they do agree that modern science only began when men started to use their own eyes instead of relying on what Aristotle and Ptolemy had written long centuries ago. Read what William Peddie has to say on " The Nature of Scientific Knowledge ".

WILLIAM PEDDIE

THE NATURE OF SCIENTIFIC KNOWLEDGE

WE come now to the question of what we are to regard as knowledge. What is the method of its attainment? What certainty have we of the truth of our knowledge? Have we to join in the wailings of the earlier Greeks that there exists no criterion; or is belief of any value to us as it was to Socrates?

Science is knowledge, and any developed body of knowledge constitutes a science whether it happens to be called by that name or not. Francis Bacon, who explicitly formulated the scientific method, made the proud boast that he had taken all knowledge as his realm. His boast was a very foolish one if it were taken literally, as he showed when he attempted to apply his own principles to astronomy, in which he was no specialist. But it was a clear-visioned one if regarded, as he meant it, as a recognition that all knowledge was subject, in its acquirement, to the method which he formulated. That is simply the old inductive method as given by Aristotle. Observations, or facts, are furnished by the senses. Reason sifts them and by means of the imagination forms a general induction, or theory, which concisely connects and describes these facts. Further, from that theory, Reason, if possible, predicts the existence of the other facts. By the verification of these predictions, the theory receives confirmation and the body of knowledge to which it belongs is added to. This is the method by which all knowledge advances. The question of its accuracy still remains. I cannot do better than state the problem in the words of Algazzali, the Arabian philosopher. In his own mental experience he passed through the historical stages of Grecian thought.

" I said to myself, ' My aim is simply to know the truth of things; consequently, it is indispensable for me to

ascertain what is knowledge '. Now, it was evident to me
that certain knowledge must be that which explains the
object to be known in such a manner that no doubt can
remain, so that in future all error and conjecture respecting
it must be impossible. Not only would the understanding
then need no efforts to be convinced of certitude, but
security against error is in such close connexion with
knowledge, that even were an apparent proof of falsehood
to be brought forward, it would cause no doubt, because
no suspicion of error would be possible. Thus, when I
have acknowledged ten to be more than three, if anyone
were to say, ' on the contrary three is more than ten, and, to
prove the truth of my assertion, I will change this rod into a
serpent ', my conviction of his error would remain unshaken.
His manœuvre would only produce in me admiration for
his ability. I should not doubt my own knowledge.

  " Then was I convinced that knowledge which I did
not possess in this manner, and respecting which I had not
this certainty, could inspire me with neither confidence
nor assurance ; and no knowledge without assurance
deserves the name of knowledge.

  " Having examined the state of my own knowledge, I
found it to be divested of all that could be said to have
these qualities unless perceptions of the senses and irre-
fragable principles were to be considered such. I then
said to myself, ' Now, having fallen into this despair, the
only hope of acquiring incontestable convictions is by the
perceptions of the senses and by necessary truths '. Their
evidence seemed to me to be indubitable. I began, however, to examine the objects of sensation and speculation,
to see if they possibly could admit of doubt. Then doubts
crowded upon me in such numbers that my incertitude
became complete. If we look at the stars, they seem to be
as small as money pieces ; but mathematical proofs convince us that they are larger than the earth. These and
other things are judged by the senses, but rejected by

reason as false. I abandoned the senses, therefore, having seen all my confidence in their truth shaken.

" ' Perhaps ' said I ' there is no assurance but in the notions of reason, that is to say, first principles, as that ten is more than three . . . to exist and not to exist at the same time is impossible.'

" Upon this the senses replied, ' What assurance have you that your confidence in reason is not of the same nature as your confidence in us ? When you relied on us, reason stepped in and gave us the lie ; had not reason been there, you would have continued to rely on us. Well, may there not exist some other judge superior to reason, who, if he appeared, would refute the judgments of reason in the same way that reason refuted us ? The non-appearance of such a judge is no proof of his non-existence.'

" I strove in vain to answer the objection, and my difficulties increased when I came to reflect on sleep. I said to myself, ' During sleep, you give to visions a reality and consistence, and you have no suspicion of their untruth. On awakening, you are made aware that they were nothing but visions. What assurance have you that all that you feel and know when you are awake does actually exist ? It is all true as respects your condition at that moment ; but it is nevertheless possible that another condition should present itself which should be to your awakened state that which your awakened state is now to sleep ; so that as respects this higher condition, your waking is but sleep.' "

Finally, Algazzali concludes that the first stage in life is that of pure sensation ; the second is that of understanding ; " the third is that of reason, by means of which the intellect perceives the necessary, the possible, the absolute, and all those higher subjects which transcend the understanding. But after this there is a fourth stage, when another eye is opened, by which man perceives things hidden from others,

perceives all that will be, perceives the things that escape
the perceptions of reason, as the objects of reason escape
the understanding, and as the objects of the understanding
escape the sensitive faculty. This is prophetism."

The last word need not surprise us if we consider what its
essential meaning must be. We would now use some such
term as " intuitive genius " instead. To prophesy is to make
an assertion for another, feeling it to be true or fit. It is the
gift of genius to make such assertions. Algazzali's " eye of
prophetism " is simply a sense of fitness. He has arrived at
the conclusion of Socrates that, if sure proof cannot be found,
belief in what is best is always possible. That is the position
of modern science. It aims at a simple, comprehensive,
accurate description ; and it adopts the fittest. To one who
has not considered the matter, it may seem astounding to be
told that science is based upon belief, but a little reflection will
make it clear. The conclusions of mathematics, the most
rigid of the supposed rigid sciences, are no more certain than
the postulates upon which they are based. Grant the postulates
and the conclusions are infallible. The postulates of Euclid,
which are the beliefs of ordinary geometry, cannot be verified,
in their applicability to our universe, as being free from an
error of about one in 10 millions.

It is not possible for us to draw, in this world, any conclusion which does not test us. And it is better so, for we
may say with Rabbi ben Ezra :—

　　" Do I remonstrate : folly wide the mark !
　　　Rather I prize the doubt,
　　　Low kinds exist without,
　　Finished and finite clods, untroubled by
　　　a spark."

We have seen there is no distinction to be drawn, as to its
ultimate nature and method of acquirement or development,

between knowledge in any branch of thought and knowledge which is usually called scientific. Therefore a consideration of the subject with which we have been dealing in this light sketch may be of equal value to each one of you, whatever be your special line of study. I have brought it before your notice in the hope that this may be so. To those of you who are students of education, which is the science of progress of knowledge in all branches, the subject is one of fundamental importance, as it is also to those who are students of science in any one of its branches. . . . There is a science of language, and the best science itself is concise, accurate, and intuitively imaginative description. So also, the truest history is a science of events and their relations. Further, I would remind those of you who aim at attainments in various directions of practice that, other things being equal, the best practitioner is the one who understands from the widest point of view. And to any of you who may intend subsequently to undertake studies in theology, I would point out a saying of Wyclif which should merit your reverence—" God bindeth not men to believe anything they cannot understand "; also this, that in founding the doctrines you teach upon fit beliefs, you are taking the same ground as science takes in her progress. She can cast no sneer at theology. The common ground of the two is their joint glory.

Each one of you must take part in the progress of knowledge ; not personal knowledge merely, but the total of human knowledge. This is a duty which no one can avoid. What then about failure ? If you wish a definition of it, take this : " A man's reach must exceed his grasp ". This means that aim must go beyond attainment. But that is also a definition of success. The climber reaches high in his search for holds ; but in order to ascend surely, he chooses the holds which are well within his power, and he often tries wrong paths before he finds the right one. A world in which there was no failure would be an imperfect world. You remember the dweller in the star Rephan, the home of perfection, where there was—

> " No hope, no fear : as to-day, shall be
> To-morrow : advance nor retreat need we
> At our standstill through eternity ! "

He asks—

> " How did it come to pass there lurked
> Somehow a seed of change that worked
> Obscure in my heart till perfection irked—"

At last a voice said to him—

> " Thou art past Rephan, thy place be Earth ".

You remember also the grammarian whose life was spent in the search for knowledge—

> " That low man seeks a little thing to do,
> 　　　Sees it and does it :
> This high man, with a great thing to pursue,
> 　　　Dies ere he knows it.
> That low man goes on adding one to one,
> 　　　His hundred's soon hit :
> This high man, aiming at a million,
> 　　　Misses an unit.
> That has the world here—should he need the next,
> 　　　Let the world mind him !
> This throws himself on God, and unperplexed
> 　　　Seeking shall find Him."

So men buried the grammarian high on the mountain-top ; for they recognized that

> " Lofty designs must close in like effects !
> 　　　Loftily lying,
> Leave him—still loftier than the world suspects,
> 　　　Living and dying."

They recognized that he had sought knowledge for its own sake.　You also may do this.

Perhaps we might usefully pursue this point a little further. *Are accepted scientific theories absolutely true?* Much depends on what we mean by truth, and it would be anticipating the Philosophy part of the Course to delve deeply into that difficulty. Meanwhile we shall find that the greatest scientists retain a great humility. Those that know most make least claim to have discovered absolute truth. It is sufficient for them if their hypotheses give a satisfactory explanation of the facts; and, if two hypotheses appear to explain them equally well, Science, following William of Occam's slogan, adopts the simpler. In the following brief extract from *The Mechanism of Nature*, Professor Andrade compares the tasks of the philosopher and the scientist, and shows how science is always ready to abandon even its most cherished theories as soon as they have served their purpose and been replaced by something a little nearer to the heart's desire of absolute, though possibly unattainable, accuracy.

E. N. DA C. ANDRADE

THE PHILOSOPHER AND THE SCIENTIST

IT is the task of the philosopher to reflect upon the general nature of the happenings, material and spiritual, that make up the life of man, and to endeavour to work out some scheme which shall help to reconcile conflicting appearances and simplify, by the investigation of first principles, the complex tangle of events in which our being is involved. It is for him to try to find some kind of an answer to the eternal " Why ? " which mankind, bewildered by the problems of good and evil, of life and death, has been uttering, now in the stammer of childhood, now in the harsh voice of agony, now in the quiet tones of reflecting age, since man has been a thinking animal.

The nature of appearance and reality, the meaning of truth and falsehood, the scope and implication of our knowledge, the significance of the conception of beauty—these are among the hard questions on which the philosopher must exercise his powers. They are very wide and very elusive, difficult to enunciate satisfactorily, more difficult to solve in the least particular. New points of view can be found, but how are we to judge when a real advance has been made ?

The task of the man of science is more modest : it is not to answer the everlasting " Why ? " but the no less everlasting " How ? " He deals entirely with the facts of observation, and tries to reduce them into a system, so that, if we admit certain principles to start with, things which are actually known to happen systematically can be shown to follow as necessary consequences, and a method of looking for new ones is suggested. The principles themselves are chosen to suit the facts, which for the man of science are all-important— *les principes ne se démontrent pas* (the principles are not open to proof). Whether the principles are in the actual sense true is not a matter on which the man of science, as such, feels called to argue : if their consequences agree with Nature they are provisionally true, at any rate.

The fundamental principles of science are, therefore, often called *working hypotheses*, since they are devised with the sole purpose of furnishing a basis upon which a system may be built which appears to correspond with the behaviour of the material world, wherever we are able to make measurements for comparison. We consider that an advance has been made when a wider range of observed phenomena has been brought within the scope of one general principle.

It follows that a scientific theory may be abandoned when it has proved itself insufficient without in any way impeaching the general validity of the scientific method. Let us consider for a moment the history of the atomic theory, as an example. Forty years ago it was generally held that atoms were hard, unbreakable entities, something like exceedingly minute billiard balls, each element possessing a perfectly definite type of atom, fundamentally different from that of any other element. This hypothesis was sufficient to explain the properties of gases, for by supposing that atoms of this kind possessed certain motions obeying the laws of mechanics, results could be deduced mathematically which agreed excellently with the properties of gases as we observe them in the laboratory. By endowing the atoms with certain forces of attraction, or affinities, we could explain the general facts of chemistry.

Then came the discovery of the electron, which is very much lighter than any atom, and suggested the possibility that the different types of atom might be built up of electrons. Further, the discovery of radio-activity showed that certain atoms, such as those of elements of the radium family, can fire off electrically charged particles, and so not only contain those particles as parts of their structure, but also must possess within themselves a store of energy, to provide the energy of the radiations. It was a question of either admitting this internal energy or of denying the principle of the conservation of energy, for the radio-active elements give out energy without any energy being put into them by us. The principle of the conservation of energy had proved too generally useful to be given up, although it was definitely suggested by some that it would have to be abandoned. The facts of radio-activity forced us, however, to abandon the idea that the atom was unbreakable, for the atoms of a radio-active element shoot off fragments of themselves and become atoms of other elements. Results of further researches could only be explained by supposing that the atom had a structure like a minute solar system, the mass of the atom being concentrated in an excessively small nucleus at the centre, and the rest of the atom consisting of electrons with wide spaces between them. . . .

Is the critic, then, justified in reproaching the physicist in this way : Forty years ago you told us that atoms were hard, indivisible, and unbreakable, made perfect in the beginning of things, and persisting in unworn perfection ever since. To-day you tell us that atoms are loose structures which can be very easily broken : you speak of radio-active atoms breaking up and changing to simpler atoms, and even speculate on the original formation of heavier atoms from lighter ones. What are we to believe ? Your accepted theories of one generation are abandoned in the next : how can I be sure that you are right this time ?

In my opinion, the correct answer is that we do not claim any absolute truth for our theories : we claim, rather, that a theory like our modern atomic theory has very great merits because all the phenomena with which we are at present acquainted are just such as we should expect if it were true. Nature, in the aspects which the physicist investigates, behaves as if there were atoms and as if they had the properties which we now claim for them. The older conception of atoms was good enough to explain the phenomena then considered, and we can still use it for certain simpler problems, where introducing the idea of atomic structure brings in needless complications ; but to explain the facts of radio-activity and of spectroscopy we must introduce the newer features of the theory.

The new theory is also better than the old because it demands only two ultimate things from which atoms are supposed to be built up ; protons and electrons. The fewer entities we need to assume as fundamental in order to explain things, the better our theory. We do not claim any finality for it : some new discovery may suddenly force us to modify our ideas in many particulars, but the successes of the present theory show that we shall probably have to retain many of the general features of the theory. It is an excellent working hypothesis because it has shown us law where law was not hitherto discovered, and connexions between different pheno-mena where before we knew of no connexion. It has enabled us to arrange our known facts in a more convenient and logical way, and has led to the discovery of very interesting new facts. It is justified by its works, but it is not final. Science is a living thing, and living things develop. . . .

On this view any particular scientific theory is a provisional tool with which we carve knowledge of the material world out of the block of Nature. It may at any moment be supplanted by a new theory, but this is only to say that when we get a better tool, which does all that this one does and something more as well, we will abandon our present tool. To refuse to use a tool because some day a better one may be invented is folly ; in the same way, not to make use of a theory which has been proved to explain a great many facts and to suggest new lines of research, because it has acknowledged flaws and is incapable of explaining other facts, would be folly. To use another metaphor, the history of science may, as has been said, be full of beautiful theories slain by ugly little facts ; but those theories did not die in vain if before their death they had subdued a vast number of jarring facts into a law-abiding populace. Nor do theories generally die a final death : often they are resurrected with some new feature which gets over the old difficulty which caused their temporary retirement.

The difference, then, between any religious belief and a scientific theory is that the former has for the believers an element of absolute truth : it is a standard by which they stand or fall, and to abandon it is dishonour and sin. The scientific theory is, however, only true as long as it is useful. The man of science regards even his best theory as a makeshift thing to help him on his way, and is always on the look-out for something better and more comprehensive.

So what we have learnt to-day may have to be unlearnt to-morrow ! What is the use, then, of our labours ?

Well, Science does not suddenly make a complete change in its teachings. What we learn to-day will help us to understand the modifications which to-morrow brings. It may indeed be essential to such understanding, and quite apart from the gain in knowledge which must follow any wide reading in the fields of science there are other benefits which may reasonably be expected to follow. Just as progress in science can only be made by the exercise of certain virtues, so any systematic reading in the subject may well foster similar virtues. *What mental rewards does the study of science bring ?* The answer is taken from Sir Archibald Geikie's *Landscape in History*. Geikie was a famous Scottish geologist whose favourite topic was the way in which different types of scenery had been formed by geological agencies. In the following address to his students he shows how enduring and valuable habits of mind are formed by scientific education.

SIR ARCHIBALD GEIKIE

THE VALUE OF ORGANIZED KNOWLEDGE

AMONG the mental habits which your education in science has helped to foster, there are a few which I would specially commend to your attention as worthy of your most sedulous care all through life.

In the first place I would place Accuracy. You have learnt in the laboratory how absolutely essential this condition is for scientific investigation. We are all supposed to make the ascertainment of the truth our chief aim, but we do not all take the same trouble to attain it. Accuracy involves labour, and every man is not gifted with an infinite capacity for taking pains. Inexactness of observation is sure, sooner or later, to be detected, and to be visited on the head of the man who commits it. If his observations are incorrect, the conclusions he has drawn from them may be vitiated. Thus all the toil he has endured in a research may be rendered of no avail, and the reputation he might have gained is not only lost but replaced by discredit. It is quite true that absolute accuracy is often unattainable ; you can only approach it.

But the greater the exertion you make to reach it, the greater will be the success of your investigations. The effort after accuracy will be transferred from your scientific work to your everyday life and becomes a habit of mind, advantageous both to yourselves and to society at large.

In the next place I would set Thoroughness, which is closely akin to accuracy. Your training here has shown you how needful it is in scientific research to adopt thorough and exhaustive methods of procedure. The conditions to be taken into account are so numerous and complex, the possible combinations so manifold, before a satisfactory conclusion can be reached. A laborious collection of facts must be made. Each supposed fact must be sifted out and weighed. The evidence must be gone over again and yet again, each link in its chain being scrupulously tested. The deduction to which the evidence may seem to point must be closely and impartially scrutinized, every other conceivable explanation of the facts being frankly and fully considered. Obviously the man whose education has inured him to the cultivation of a mental habit of the kind is admirably equipped for success in any walk of life which he may be called upon to enter. The accuracy and thoroughness which you have learnt to appreciate and practise at college must never be dropped in later years. Carry them with you as watchwords, and make them characteristic of all your undertakings.

In the third place we may take Breadth. At the outset of your scientific education you were doubtless profoundly impressed by the multiplicity of detail which met your eye in every department of natural knowledge. When you entered upon the study of one of these departments you felt, perhaps, almost overpowered and bewildered by the vast mass of facts with which you had to make acquaintance. And yet as your training advanced, you gradually came to see that the infinite variety of phenomena could all be marshalled, according to definite laws, into groups and series. You were led to look beyond the details to the great principles that underlie them and bind them into a harmonious and organic whole. With the help of a guiding system of classification, you were able to see the connexion between the separate facts, to arrange them according to their mutual relations, and thus to ascend to the great general laws under which the material world has been constructed. With all attainable thoroughness in the mastery of detail, you have been taught to combine a breadth of treatment which enables you to find, and keep, a leading clue even through the midst of what might seem a tangled web of confusion. There are some men who cannot see the wood for trees, and who consequently can never attain great success in scientific investigation. Let it be your aim to master fully the details of the tree, and yet to maintain such a breadth of vision as will enable you to embrace the whole forest within your ken. I need not enlarge on the practical value of this mental habit in everyday life, nor point out the excellent manner in which a scientific education tends to develop it.

In the fourth place, I would inculcate the habit of wide Reading in scientific literature. Although the progress of science is now too rapid for any man to keep pace with the advance in all its departments, you should try to hold yourselves in touch with at least the main results arrived at in other branches than your own ; while, in that branch itself, it should be your constant aim to watch every onward step that is taken by others, and not to fall behind the van. This task you will find to be no light one. Even if it were confined to a survey of the march of science in your own country, it would be arduous enough to engage much of your time. But science belongs to no country, and it continues its onward advance all over the globe. If you would keep yourselves informed regarding this progress in other countries, as you are bound to do if you would not willingly be left behind, you will need to follow the scientific literature of those countries. . . .

In the fifth place, let me plead for the virtue of Patience. In a scientific career we encounter two dangers for the avoidance of which patience is our best support and guide. When life is young and enthusiasm is boundless ; when from the details which we may have laboriously gathered together we seem to catch sight of some new fact or principle, some addition of more or less importance to the sum of human knowledge, there may come upon us the eager desire to make our discovery known. We may long to be allowed to add our own little store to the growing temple of science. We may think of the pride with which we should see our names enrolled among those of the illustrious builders by whom this temple has been slowly reared since the infancy of mankind. So we commit our observations to writing and send them to publication. Eventually we obtain the deep gratification of appearing in print among well-known authors in science. Far be it from me to condemn this natural desire for publicity. But, as your experience grows, you will probably come to agree with me that if the desire were more frequently and energetically curbed, scientific literature would gain much thereby. There is amongst us far too much hurry in publication. We are so afraid lest our observations should be forestalled—so anxious not to lose our claim to priority, that we rush before the world often with a half-finished performance which must be corrected, supplemented, or cancelled by some later communication. It is this feverish haste which is largely answerable for the mass of jejune, ill-digested, and erroneous matter that cumbers the pages of modern scientific journals. Here it is that you specially need patience. Before you venture to publish anything, take the utmost pains to satisfy yourself that it is true, that it is new, and that it is worth putting into print. And be assured that this reticence, while it is a kindness to the literature of science, will most certainly bring with it its own reward to yourselves. It will increase your confidence, and make your ultimate contributions more exact in their facts, as well as more accurate and convincing in their argument.

The other danger to which I referred as demanding patience is of an opposite kind. As we advance in our career, and the facts of our investigations accumulate around us, there will come times of depression when we seem lost in a labyrinth of detail out of which no path appears to be discoverable. We have, perhaps, groped our way through this maze, following now one clue, now another, that seemed to promise some outlet to the light. But the darkness has only closed around us the deeper, and we feel inclined to abandon the research as one in which success is, for us at least, unattainable. When this blankness of despair shall come upon you, take courage under it, by remembering that a patient study of any department of

Nature is never labour thrown away. Every accurate observation you have made, every new fact you have established, is a gain to science. You may not for a time see the meaning of these observations, nor the connexion of these facts. But their meaning and connexion are sure in the end to be made out. You have gone through the labour necessary for the ascertainment of truth, and if you patiently and watchfully bide your time, the discovery of the truth itself may reward your endurance and your toil.

It is by failures as well as by successes that the true ideal of the man of science is reached. The task allotted to him in life is one of the noblest that can be undertaken. It is his to penetrate into the secrets of Nature, to push back the circumference of darkness that surrounds us. To disclose ever more and more of the limitless beauty, harmonious order, and imperious laws that extend throughout the universe. And while he thus enlarges our knowledge, he shows us also how Nature may be made to minister in an ever augmenting multiplicity of ways to the service of humanity. It is to him and his conquests that the material progress of our race is mainly due. If he were content merely to look back over the realms which he has subdued, he might well indulge in jubilant feelings, for his peaceful victories have done more for the enlightenment and progress of mankind than were ever achieved by the triumphs of war. But his eye is turned rather to the future than to the past. In front of him rises the wall of darkness that shrouds from him the still unknown. What he has painfully accomplished seems to him but little in comparison with the infinite possibilities that lie beyond. And so he presses onward, not self-satisfied and exultant, but rather humbled and reverential, yet full of hope and courage for the work of further conquest that lies before him.

BOOK II
THE UNIVERSE

THE UNIVERSE

IN this book we shall trace the development of man's ideas about the universe, and the place in it of the earth on which he lives. It was probably the regularity of day and night, of the waxing and the waning of the moon, and of the recurrence of the seasons which first suggested to him the idea of *order*, and of natural laws which would explain this uniformity and discipline. First let us glance at some of the early ideas of the Greeks. We shall find them somewhat naïve, but full of interest, especially if we try to see what can be said for them as first attempts at explanation.

How did the Greeks explain the composition of matter? For the answer we go to Pythagoras, who lived about half a century after Thales, and who is generally given the credit for the first proof of the theorem about the square on the hypotenuse of a right-angled triangle. According to Ovid, Pythagoras " examined all things with his mind,` and with watchful study, and then gave them to be learned by the public ; and he sought the crowds of people as they sat in silence, and wondered at the revealed origin of the vast universe, and what was the cause of things, and what Nature meant, and what was God ". The following brief summary of the Greek theory of the four elements is put into the mouth of Pythagoras by Ovid.

PYTHAGORAS

THE FOUR ELEMENTS

THE everlasting universe contains but four elementary substances. Two of these, namely earth and water, are heavy. They are borne downwards by their weight. The other two—air, and fire which is still purer than air—nothing pressing them, seek the higher regions. Although these are separated in space, yet all things are made from them, and are resolved into them. The earth, dissolving, distils into flowing water ; water, evaporating, departs in the breezes and changes into

45

air ; the most subtle air, its weight being removed again, shoots upwards into the fires on high. Then the order is reversed and they return back again ; for fire, becoming gross, passes into dense air ; this changes into water ; and earth is formed of water made dense.

No form remains for ever, and Nature, the renewer of all things, makes one shape grow into another. In this vast universe, believe me, nothing perishes ; it just alters and changes its appearance. And to begin to be something different from before is called being born ; and to cease to be the same thing is to die. Things can be brought here and carried there, yet they never cease to exist.

My own belief is that nothing lasts long under the same form. Even the golden age changed to the iron age, and no place endures unchanged. I have seen the sea cover what had once been solid dry land, and I have seen the land encroach on the sea. Far away from the ocean we find sea-shells lying, and old anchors are found on the tops of mountains. That which was a plain has been hollowed out by flowing water into a valley ; and by a flood a mountain has been levelled into a plain. Ground that was swampy has become desert ; and places long dusty and dry are now wet with standing pools. Here nature has caused fresh springs to flow, while there she has dried them up ; and rivers have burst forth, aroused by ancient earthquakes in one place, while elsewhere they have subsided and vanished.

Every schoolboy smiles at this theory now, but it is not difficult to see why the Greeks accepted it. They knew nothing of gravitation—two thousand years were to roll by before Newton came—and the theory did explain why flames soared upwards while solid objects fell downwards. It did suggest that stones sank in water because they were the most " earthy " form of matter, while wood, having more air and water in its composition, could float. Consider again how corn and olive trees and grass and every other kind of plant all grow in the same soil ; surely

" earth " must be an important constituent of them all. And when wood or straw or dry grass burns, the same sort of flame comes out of them all, so that they must all have " fire " in their composition. Even to-day if we should happen to be caught in a violent thunderstorm without any shelter we would probably begin to think that the old theory was not so very wide of the mark.

The two most important questions for the Greeks, as for us, dealt with Man's place in nature, and with the place of the Earth in the universe. The brilliant philosophers who attended Plato's Academy must have spent much time in trying to devise a rational explanation of the movements of the sun and moon, stars and planets. The problem of the source of this motion—what Aristotle called " the Efficient Cause "—constantly exercised their minds. *What was the Greek view of the Universe?* Aristotle has preserved this for us, and in his book *On the Heavens* from which the following extracts are taken he gives us not only his own views, but those of other philosophers. It will be seen that even in those early days there were people more anxious to " adjust the facts " to fit their own preconceived theories than to make their opinions harmonize with the observed facts.

ARISTOTLE

ON THE HEAVENS

ALL men believe that there are gods, and all men, both barbarians and Greeks, assign the highest place in heaven to the divine nature. . . . For, according to tradition, in the whole of past time no change has taken place either in the heaven as a whole or in any of its parts. Moreover, the name by which we call it appears to have been handed down in succession from the ancients, who held the same opinion about its divine nature which lasts to the present time.

For such reasons, then, we believe that the heaven was neither created nor is it corruptible, but that it is one and everlasting, unchanged through infinite time. Hence we may well persuade ourselves that ancient assertions, especially those of our own ancestors, are true, and see that one kind of motion is immortal and divine, having no end, but being itself the end of other motions. Now motion in a circle is perfect, having neither beginning nor end, nor ceasing in infinite time.

As the ancients attributed heaven and the space above it to the gods, so our reasoning shows that it is incorruptible and uncreated and untouched by mortal troubles. No force is needed to keep the heaven moving, or to prevent it moving in another manner . . . nor need we suppose that its stability depends on its support by a certain giant Atlas, as in the ancient fable : as though forsooth all bodies on high possessed gravity and an earthly nature. Not thus has it been preserved for so long nor yet, as Empedocles asserts, by whirling round faster than its natural motion downwards. Nor is it reasonable to think that it remains unchanged by the compulsion of a soul, untiring and sleepless, unlike the soul of mortal animals, for it would need the fate of some Ixion (bound for ever to a fiery wheel) to keep it in motion. . . .

The heaven, moreover, must be a sphere, for this is the only form worthy of its essence, as it holds the first place in nature. . . . Every plane figure is contained by straight lines or by a circumference. The right-lined figure is bounded by many lines, but the circle by but one. But as the one is prior to the many and the simple to the composite, so the circle is the first of plane figures. . . . Again, to a straight line an addition can always be made, but to a circular line never. Thus once more the line which traces a circle is perfect. Hence if the perfect is prior to the imperfect, the circle again will be the first of figures. In like manner also, the sphere will be the first of solids ; for this alone is contained by one superficies, while flat-sided figures are contained by many. As a circle is in planes, so is a sphere in solids. . . .

Further still, since it seems clear and we assume that the universe revolves in a circle, and since beyond the uttermost sky is neither body nor space nor vacuum, once more it follows that the universe is spherical. For, if it were rectilinear, there must be space beyond it : a rectilinear body as it revolves will never occupy the same place : where it formerly was it is not now, and where it is not now it will be again, because the corners project. . . .

It remains to discuss the earth—where it is situated, whether it is at rest or moves, and what is its form. With regard to its position, all philosophers have not the same opinion. Most of those who assert that the heaven is finite say that the earth lies at the centre, while those in Italy who are called Pythagoreans hold the contrary. For they say that at the centre of the universe is fire, and that the earth, being one of the stars, moves in a circle about that centre and thus causes day and night. They also invent opposite to our earth another earth, which they call the counter-earth : not investigating theories and causes to explain the facts, but adjusting the facts to fit certain opinions and theories of their own. To many others also it seems that a central place should not be assigned to the earth for reasons not based on facts but on opinions. For they fancy that the most honourable place belongs to the most honourable nature : that fire is more honourable than earth and the boundaries of a space than the region within. But the circumference and the centre they say are boundaries. So that, thus reasoning, they think not the earth but fire holds place in the centre of the sphere. Further still, the Pythagoreans hold that the chief place should be best guarded, and call the centre the altar of Zeus, and thus again assign this place to fire as if the centre of a mathematical figure and the middle of a thing or the natural centre were of the same kind. . . . Such as assert that the earth is not situated in the middle of the universe are of opinion that it and the counter-earth also move round the centre in a circle. And to some it appears that many such bodies may move round the centre though invisible to us by the intervention of the earth. Hence they say there are more eclipses of the moon than of the sun, for each of the moving bodies, and not the earth only, can obstruct the light of the moon. . . . But some say that the earth, being

E

situated in the centre, rolls round the pole which is extended through the universe, as it is written in the *Timæus*.

In a similar way there is doubt about the shape of the earth. To some it seems to be spherical, but to others flat, in the form of a drum. To support this opinion they urge that, when the sun rises and sets, he appears to make a straight and not a circular occultation, as it should be if the earth were spherical. These men do not realize the distance of the sun from the earth and the magnitude of the circumference, nor do they consider that, when seen cutting a small circle, a part of the large circle appears at a distance as a straight line. Because of this appearance, therefore, they ought not to deny that the earth is round. . . . It is indeed irrational not to wonder how it is that a small fragment of earth if dropt from a high place moves downward and a larger fragment more swiftly downward, while the whole earth does not tend downward and its great bulk is at rest. For if, while fragments of earth are falling, some one could take away the whole earth before they reached it, they would nevertheless move downward if nothing opposed them. Hence this question is of general philosophic interest, its consequences seeming no less difficult than the problem. For some on this account hold that the part of the earth below us must be infinite, as Xenophanes of Colophon says, rooted to infinity. . . . Hence the rebuke of Empedocles when he writes :—

> " The boundless depths of earth, the æther vast,
> In vain the tongues of multitudes extol
> Who see but little of the mighty all."

But others say the earth floats upon water. This view we consider the most ancient : it is ascribed to Thales the Milesian. It regards the earth as upheld in its place because it floats like a piece of wood or anything else of the same kind. . . . But water itself cannot remain suspended on high, but must be upheld in its turn by something. Further, as air is lighter than water, so water is lighter than earth. How then can they fancy that what is lighter lies below and supports what is heavier ? Again, were the whole earth to float upon water, this would also be the case with its fragments. But this seems not so, for any piece of earth sinks to the bottom of water, and larger fragments sink more swiftly.

Aristotle's cosmology was far from being on the same high level as his biology, and in the next century he was outstripped by Aristarchus of Samos, who first placed the sun at the centre of the universe, and suggested that the Earth and the other planets revolve round it. The book in which he outlined this theory has been lost, but we know that it existed from a clear reference to it in the works of Archimedes. Aristarchus, however, found few to agree with him. It was only natural that the Greeks should consider Man as the most important being in the universe (after God, or the gods) and the Earth as the most important body, and therefore the fixed centre of that universe. Hipparchus gave a mathematical explanation of a universe which revolved round the Earth in cycles and epicycles, and Ptolemy of Alexandria, about A.D. 127 to 151, expounded and elaborated this theory.

Ptolemy said that each of the heavenly bodies revolved round the earth in its sphere, but also revolved in its own little epicycle in this sphere. Thus the somewhat erratic movements of the planets could be fairly satisfactorily explained. The moon was in the sphere nearest to the earth ; then came Mercury, then Venus, then the Sun, Mars, and Jupiter in that order, with Saturn farthest away of the planets, and the sphere of the fixed stars outside all.

With Ptolemy, and with his successors for many centuries, astronomy and astrology were one. A belief in astrology was almost universal all through the Dark Ages and mediæval times. It coloured all other sciences, and we can see its influence very strongly in medicine, in botany, and in chemistry, or rather alchemy. *How were the stars supposed to influence human affairs ?* The answer is given in this brief extract from Ptolemy's *Treatise in Four Books*, usually called his " Tetrabiblos " or his " Quadripartitum ".

PTOLEMY

THE INFLUENCE OF THE STARS

A LITTLE thought should make it clear that a certain power comes from the everlasting and unchanging ethereal element, and suffuses the whole region of the earth, where changes are always occurring. Of the four primary sublunar elements, fire and air are affected by the motions in the ether, and these in turn affect all other things—earth and water and all living things.

We can easily see the continual effect of the sun on everything on the earth ; not only in causing the seasons, but in making animals breed, and plants produce their crops ; in making the rivers flow, and in producing changes in bodies. And its daily revolution round the earth provides heat or cold, moisture or dryness, in regular order, corresponding with its height towards the zenith.

The moon, too, since it is the heavenly body nearest the earth, has a most important effect on all mundane things, for most of them, whether living or not, are sympathetic to her, and change as she changes. The flow of rivers increases and decreases with her light ; the tides of the sea turn with her rising and setting ; and plants and animals, in some part if not as a whole, wax and wane with her.

Moreover the passages of the fixed stars and the planets across the sky often mean hot, windy, or snowy conditions in the atmosphere, and things on the earth are affected correspondingly. Then, too, their aspects to one another, whether in opposition, trine, quartile, or sextile positions, by the meeting and mingling of their influences, can cause many complicated changes. For though the sun's power is in general supreme, the other heavenly bodies can aid or oppose it in particular details, the moon more obviously and continuously, especially when it is new, at quarter, or full, and the stars in a more subtle way, and only on occasions when they are so placed as to exert their full power.

The sun's active power produces a heating and to some extent a drying effect. We can see this power at work more easily in the case of the sun because it is so much bigger than the other heavenly bodies, and because its effects are so noticeable in the changes of the seasons, for the closer it approaches to the zenith the greater its heat.

The moon is closer to the earth, and because the moist exhalations from the earth nourish it most of all the heavenly bodies, its chief effect is in humidifying, and in softening and causing putrefaction in bodies. However, it shares to some extent in the sun's heating power because of the light which it receives from the sun.

The star of Saturn is farthest removed from the sun's heat and from the earth's moist vapours. Its effect therefore is chiefly to cool and to some extent to dry. . . .

The star of Mars is mainly powerful to dry and to burn, as might be expected from its fiery colour, and its nearness to the sun, for the sun's sphere lies just below that of Mars.

The star of Jupiter has a temperate power because his movement takes place between the cooling Saturn and the burning Mars. He brings both heat and moisture, and because his heating power is the greater because of the underlying spheres he produces fertilizing winds.

The star of Venus has the same powers and the temperate nature of Jupiter, but the effect is different because with Venus the moisture is greater than the heat, as with the moon. She warms because of her nearness to the sun, but even greater is the humidifying power because of her own light and her gathering up of the moisture in the earth's atmosphere.

Mercury is sometimes very drying because he is never far removed in longitude from the heat of the sun. At other times he is humidifying because his sphere is next above that of the moon, which is closest to the earth. And he changes quickly from one nature to the other, inspired by the speed of his own motion in the neighbourhood of the sun itself.

[Jupiter, Venus, and the Moon are beneficent because of their temperate nature, and because of their heat and moisture. Saturn and Mars bring evil because of the excessive coldness of the former and the dryness of the latter. The Sun and Mercury are sometimes good, sometimes bad. The Sun, Saturn, Jupiter, and Mars are masculine ; the Moon and Venus are moist and therefore feminine. Mercury is masculine as a morning star and feminine as an evening one, and in general the morning stars are masculine and the evening stars feminine. The nature of the larger fixed stars is also given.]

From all this it will be evident that the power of each of the planets must be considered not only from its own natural character, but also with reference to the sign of the Zodiac in which it appears. Their power is greater when they are oriental, and adding to their own proper motion. If they are occidental their speed with respect to the earth is diminishing, and then their energy will be less. Their position relative to the horizon is also important ; for they are most powerful when they are in mid heaven, or approaching it, and next most powerful when on the horizon, and thirdly when 30 degrees above it. . . .

In general the moon's influence is most powerful during the first four years of life : then Mercury takes over for the next ten years. Venus dominates us from the age of 14 to 22, and then the sun begins a long period of supremacy lasting for nineteen years. Mars then rules for fifteen years, and Jupiter follows for 12, only to give way to Saturn, who remains the dominant influence for the last years of our lives.

The knowledge of the ancients was preserved for us in the works of Aristotle and Ptolemy, although even these were almost completely forgotten throughout the Dark Ages. A few Arabic translations were kept in circulation, however, and the Arabs brought them to Spain and Sicily. In the twelfth century a few scholars, hungry for knowledge in an age of faith, travelled from other countries of Europe to Spain and Sicily and Syria in order to learn Arabic and translate these Arabic versions into Latin. One of the earliest and greatest of these scholars was Adelard of Bath, to whom we owe the introduction of Greek mathematics and Ptolemy's astronomy into the West. It was these Latin translations from the Arabic which stimulated a desire for the Greek originals, and for Latin translations from them, and with the spread of these came the flood tide of the Renaissance.

The philosophy of Aristotle and the astronomy of Ptolemy were harmonized with Christian dogma by Thomas Aquinas, and for two or three centuries the Church authorities forbade any questioning of their theories or their statements. Science lay dormant. *Who first awakened cosmology from its fourteen-hundred years' sleep?* The answer is given in the following extracts from a famous book which Copernicus published just before he died in 1543.

NICOLAUS COPERNICUS

THE REVOLUTION OF THE CELESTIAL GLOBE

THE ancient philosophers affirm the earth to be at rest at the centre of the universe, and doubtless to maintain itself there. But, if anyone holds that the earth revolves, he affirms at any rate that the motion is natural and not forced. Which things indeed are according to nature, for motions which are forced produce different effects.

It is true that things to which force or impact is applied necessarily disintegrate, and cannot hold together for long; but those which have a natural motion maintain themselves steadily, and are preserved in coherence. It is without reason, therefore, that Ptolemy fears lest the earth, if it move, should disintegrate and all terrestrial things be thrown into confusion by the power of nature, so far beyond that of art, or anything possible to human ingenuity. But why, on his view of a fixed earth and a moving sky, does he not fear it even more for the sky, whose motion is so much the more swift, as the heaven is greater than the earth ? Is the sky perhaps an immense structure, which by the power of an ineffable motion is separated from its centre, and on the other hand falls in ruin if it stand still ? Surely, if this be so, the magnitude of the heaven will increase to infinity. For by how much the further from the centre the motion is carried by its own impetus, by so much the motion will be swifter, on account of the ever-growing circumference which it must describe in order to traverse space in twenty-four hours : and, in turn, the immensity of the heaven grows greater by reason of this increasing motion. Thus the velocity increases the magnitude and the magnitude the velocity to infinity.

And next comes the physical axiom : *That which is infinite is unable to travel forwards, nor can it move through any cause.* The heaven, therefore, necessarily stands still. But, they say, beyond the sky is neither body, nor space, nor vacuum—in a word nothing, and therefore no means exist of breaking out through the sky ; then indeed it is wonderful, if from nothing anything is able to hold together. And if the heaven were infinite, and only bounded by its inner concavity, all the more can it be proved, perhaps, that there is nothing beyond the sky ; since each and everything is in it, whatever space it may occupy, the heaven will remain immoveable. . . . Whether therefore the universe be finite, or infinite, let us put aside the disputations of the physicists, having this certain—that the earth to its poles is bounded by a closed, spherical surface. Why, therefore, do we hesitate to concede to it an appropriate mobility, rather than that the whole universe should fall in ruin, the end of which is unknown and unknowable ; or confess of the daily revolution itself that in the heaven it is an appearance but in the earth reality ? And this is indeed true, as Virgil's Æneas affirms when he says

" As I am carried away from the harbour both town and shore recede from my view."

[And here is Copernicus's idea of the universe.]

First and above all lies the sphere of the fixed stars, containing itself and all things, for that very reason immoveable ; in truth the frame of the universe, to which the motion and position of all other stars are referred. Though some men think it to move in some way, we assign another reason why it appears to do so in our theory of the movement of the earth. Of the moving bodies first comes Saturn, who completes his circuit in thirty years. After him, Jupiter, moving in a twelve-year revolution. Then Mars, who revolves biennially. Fourth in order an annual cycle takes place, in which we have said is contained the earth, with the lunar orbit as an epicycle. In the fifth place Venus is carried round in nine months. Then Mercury holds the sixth place, circulating in the space of eighty days. In the middle of all dwells the Sun. Who, indeed, in this most beautiful temple would place the torch in any other or better place than one whence it can illuminate the whole at the same time ? Not inaptly, some call it the lamp of the universe, others its mind, others again its ruler—Trismegistus, the visible God, Sophocles' Electra, the contemplation of all things. And thus rightly in as much as the Sun, sitting on a royal throne, governs the circumambient family of stars. . . . We find, therefore, under this orderly arrangement, a wonderful symmetry in the universe, and a definite relation of harmony in the motion and magnitude of the orbs, of a kind it is not possible to obtain in any other way.

Half a century of yawning and stretching and dozing again is not unreasonable after fourteen hundred years of sleep, and science certainly indulged in this after the publication of Copernicus's book. Then a young Florentine with a penetrating and enquiring mind began to doubt the teachings of Aristotle and to make experiments which justified his doubts. Aristotle had said that a heavier stone would naturally fall faster than a light one. One day in 1590 young Galileo Galilei climbed the hundred and eighty feet to the top of the leaning tower at Pisa, and dropped two balls, one heavy and one light. They fell side by side and reached the ground together. Some of the spectators were so sure that the heavier ball would fall more quickly that they declared that it had done so, and that they had seen it do so with their own eyes. But Galileo knew better. Still, he admitted that objects falling freely did move rather too fast for accurate measurement of their velocities. So he rolled the balls down inclined planes of various angles. And on the result of these experiments the foundation of the science of dynamics was laid.

Galileo showed that the motion of the rolling or falling body was steadily accelerated by the same amount every second. He showed that a ball rolling down one plane would run up another to an equal height whatever the inclination of the second plane might be. And from the fact that when the second plane was made horizontal the ball would go on rolling until brought to rest by friction he reached the idea of " inertia "—that a body tends to maintain its state either of rest or of uniform motion in a straight line. Aristotle felt that an " unmoved mover " was necessary to keep any body in motion. Galileo showed this was unnecessary.

Copernicus had given a simple explanation of the movements of the heavenly bodies with his heliocentric theory. Galileo's dynamics enabled further simplifications to be made. And he found some most valuable supporting material for the new ideas in the marvels which he saw through his home-made telescopes. *What wonderful things was Galileo the first of all men to see ?* The following extracts are taken from his *Sidereal Messenger*, published in 1610.

GALILEO GALILEI

THE SURFACE OF THE MOON

ABOUT ten months ago a report came to my ears that a Dutchman had constructed a perspective glass by the aid of which objects at a great distance from the eye of the observer were seen distinctly as if close at hand. Some proofs of the wonderful performances of this instrument were reported, and some people accepted them as true, but others would not believe them.

A few days later I received confirmation of the reports in

a letter from Jacques Badovere, a French nobleman in Paris, and this caused me to make up my mind to find out how this telescope worked, and then to make a similar instrument for myself. After a little while I succeeded in doing this, thanks to a close study of the theory of Refraction. So I made a tube, at first of lead, in the two ends of which I fitted glass lenses, both flat on one side, but on the other side one was spherically convex and the other concave. Then, bringing my eye to the concave lens, I was delighted to see objects seem much larger and nearer, for they appeared one-third of the distance away, and nine times larger than when viewed with the naked eye alone.

Soon afterwards I constructed another telescope with greater care, and this magnified objects more than sixty times. At length, by sparing neither effort nor expense, I succeeded in making for myself an instrument so much better that objects seen through it appear magnified nearly a thousand times, and more than thirty times nearer than if viewed with the unaided human eye.

It would be altogether a waste of time to enumerate the number and the importance of the benefits which will be derived from the use of the telescope, both by land and by sea. But without worrying about terrestrial objects I betook myself to observations of the heavenly bodies; and first of all I viewed the Moon as near as if it were scarcely two semi-diameters of the Earth distant. After the Moon I turned again and again to the observation of other heavenly bodies, both fixed stars and planets, with incredible delight. . . .

Let me speak first of the surface of the Moon which is turned towards us. Two areas are easily distinguished, a brighter and a darker. The bright part seems to surround and pervade the whole full Moon, while the shaded part, like a sort of cloud, discolours the surface, and makes it appear covered with spots. Now these spots, being rather dark, and of great size, are plain to everyone, and men in all ages have seen them. So I shall call them great or ancient spots. This will distinguish them from other spots, smaller in size, but so thickly scattered that they seem to be sprinkled over the whole surface of the Moon, and especially over the brighter portion of it.

These spots have never been observed by anyone before me, and from oft-repeated examination of them I have been led to the opinion which I have already expressed, namely that I feel sure the surface of the Moon is not perfectly smooth, even, and exactly spherical, as so many philosophers insist that it and all the other heavenly bodies should be. On the contrary it is full of inequalities, uneven, full of hollows and protuberances, just like the surface of the Earth itself, varied everywhere by lofty mountains and deep valleys.

Here are the observations which support this conclusion. On the fourth or fifth day after New Moon, when we see a crescent in the sky, the line dividing the sunlit portion from the part still in the shadow does not extend continuously in a smooth curve, as would happen with a perfectly spherical body. It is marked out by an irregular, uneven, and very wavy line. Some bright spots can be seen in the otherwise dark part, and patches of shadow remain in the light. Nay, a great number of small dark spots, quite separated from the dark part, still remain sprinkled almost everywhere over the area which is flooded with the Sun's light—except alone on that part of the surface which is occupied by the great or ancient spots. And I have noticed that these small dark spots just mentioned always have the dark part towards the Sun's position, while on the side away from the sun they have brighter boundaries, as if they were crowned with shining summits.

Now we have an appearance quite similar on the Earth about sunrise, when the valleys are still in shadow, while the mountains surrounding them on the side opposite to the Sun are already ablaze with the splendour of his beams. And just as the shadows in the hollows of the Earth diminish in size as the Sun rises higher, so also these spots on the Moon lose their blackness as the sunlit part grows larger and larger.

Again, not only is the boundary between light and shadow on the Moon uneven and wavy. Still more astonishing is the fact that many bright points, quite separate and apart from the illuminated portion, appear within the dark area of the moon, and at some considerable distance inside it too. After a little while these gradually increase in size and brightness, and after an hour or two they become joined to the bright area, which has spread to include them. And meanwhile other bright points, one here and another there, shooting up as if they were growing, are lit up in the shaded portion, increase in size, and at last are joined to the same illuminated surface, now still larger.

Is it not the case on the Earth before sunrise that while the level plain is still in shadow, the peaks of the most lofty mountains are illuminated by the Sun's rays ? After a little while does not the light spread further, as the middle and larger areas of those mountains are becoming sunlit ? And at length, when the sun has risen, do not the illuminated parts of the plains and hills join together ? The grandeur, however, of such peaks and valleys in the Moon seems to far surpass the ruggedness of the Earth's surface, as I shall hereafter show.

Aristotle had said that all the heavenly bodies must be perfect spheres, so that Galileo's assertion that the moon's surface was covered with irregularities at once aroused suspicious opposition among the authorities. But his next discovery was even more disturbing. It was that of four celestial bodies which quite obviously did *not* revolve round the earth.

GALILEO GALILEI

THE DISCOVERY OF JUPITER'S SATELLITES

I HAVE now finished my brief account of the observations which I have thus far made with regard to the Moon, the Fixed Stars, and the Galaxy. There remains the matter, which seems to me to deserve to be considered the most important in this work, namely, that I should disclose and publish to the world the occasion of discovering and observing the four PLANETS, never seen from the very beginning of the world up to our own times, their positions, and the observations made during the last two months about their movements and their changes of magnitude; and I summon all astronomers to apply themselves to examine and determine their periodic times, which it has not been permitted me to achieve up to this day, owing to the restriction of my time. I give them warning, however, again, so that they may not approach such an inquiry to no purpose, that they will want a very accurate telescope, and such as I have described in the beginning of this account.

On the 7th day of January in the present year, 1610, in the first hour after sunset, when I was viewing the constellations of the heavens through a telescope, the planet Jupiter presented itself to my view, and as I had prepared for myself a very excellent instrument, I noticed a circumstance which I had never been able to notice before, owing to want of power in my other telescope, namely, that three little stars, small but very bright, were near the planet ; and although I believed them to belong to the number of fixed stars, yet they made me somewhat wonder, because they seemed to be arranged exactly in a straight line, parallel to the ecliptic, and to be brighter than the rest of the stars, equal to them in magnitude. The position of them with reference to one another and to Jupiter was as follows :—

East   *   *   O   *   West

On the east side there were two stars, and a single one towards the west. The star which was furthest towards the east, and the western star, appeared rather larger than the third.

I scarcely troubled at all about the distance between them and Jupiter, for, as I have already said, at first I believed them to be fixed stars ; but when on January 8, led by some fatality,

I turned again to look at the same part of the heavens, I found a very different state of things, for there were three little stars all west of Jupiter, and nearer together than on the previous night, and they were separated from one another by equal intervals, as the accompanying figure shows.

East   O   *   *   *   West

At this point, although I had not turned my thoughts at all upon the approximation of the stars to one another, yet my surprise began to be excited, how Jupiter could one day be found to the east of all the aforesaid fixed stars when the day before it had been west of two of them ; and forthwith I became afraid lest the planet might have moved differently from the calculation of astronomers, and so had passed those stars by its own proper motion. I therefore waited for the next night with the most intense longing, but I was disappointed of my hope, for the sky was covered with clouds in every direction.

But on January 10 the stars appeared in the following position with regard to Jupiter, the third, as I thought, being hidden by the planet.

East   *   *   O   West

They were situated just as before, exactly in the same straight line with Jupiter, and along the Zodiac.

When I had seen these phenomena, as I knew that corresponding changes of position could not by any means belong to Jupiter, and as, moreover, I perceived that the stars which I saw had always been the same, for there were no others either in front or behind, within a great distance, along the Zodiac,—at length, changing from doubt into surprise, I discovered that the interchange of position which I saw belonged not to Jupiter, but to the stars to which my attention had been drawn, and I thought therefore that they ought to be observed henceforward with more attention and precision.

Accordingly, on January 11 I saw an arrangement of the following kind :—

East   .*   *   O   West

namely, only two stars to the east of Jupiter, the nearer of which was distant from Jupiter three times as far as from the star further to the east ; and the star furthest to the east was nearly twice as large as the other one ; whereas on the previous night they had appeared nearly of equal magnitude. I therefore concluded, and decided unhesitatingly, that there are three stars in the heavens moving about Jupiter, as Venus and Mercury round the Sun ; which at length was established as clear as daylight by numerous other subsequent observations. These observations also established that there are not only three, but four, erratic sidereal bodies performing their revolutions round Jupiter, observations of whose changes of position made with more exactness on succeeding nights the following account will supply. I have measured also the intervals between them with the telescope in the manner already explained. Besides this, I have given the times of observation, especially when several were made in the same night, for the revolutions of these planets are so swift that an observer may generally get differences of position every hour.

Jan. 12.—At the first hour of the next night I saw those heavenly bodies arranged in this manner :—

East   *   * O   ✳   West

The satellite furthest to the east was greater than the satellite furthest to the west ; but both were very conspicuous and bright ; the distance of each one from Jupiter was 2 minutes. A third satellite, certainly not in view before, began to appear at the third hour : it nearly touched Jupiter on the east side, and was exceedingly small. They were all arranged in the same straight line, along the ecliptic.

Jan. 13.—For the first time four satellites were in view, of the following position with regard to Jupiter :—

East   *   O   * * *   West

There were three to the west, and one to the east; they made a straight line nearly, but the middle satellite of those to the west deviated a little from the straight line towards the north. The satellite furthest to the east was at a distance of 2′ from Jupiter; there were intervals of 1′ only between Jupiter and the nearest satellite, and between the satellites themselves, west of Jupiter. All the satellites appeared of the same size, and though small they were very brilliant, and far outshone the fixed stars of the same magnitude. . . .

These are my observations upon the four Medicean planets, recently discovered for the first time by me; and although it is not yet permitted me to deduce by calculation from these observations the orbits of these bodies, yet I may be allowed to make some statements, based upon them, well worthy of attention.

And, in the first place, since they are sometimes behind, sometimes before Jupiter, at like distances, and withdraw from this planet towards the east and towards the west only within very narrow limits of divergence. and since they accompany this planet alike when its motion is retrograde and direct, it can be a matter of doubt to no one that they perform their revolutions about this planet, while at the same time they all accomplish together orbits of twelve years' length about the centre of the world.

Moreover, they revolve in unequal circles, which is evidently the conclusion to be drawn from the fact that I have never been permitted to see two satellites in conjunction when their distance from Jupiter was great, whereas near Jupiter two, three, and sometimes all four, have been found closely packed together. Moreover, it may be detected that the revolutions of the satellites which describe the smallest circles round Jupiter are the most rapid, for the satellites nearest to Jupiter are often to be seen in the east, when the day before they have appeared in the west, and contrariwise. Also the

F

satellite moving in the greatest orbit seems to me, after carefully weighing the occasions of its returning to positions previously noticed, to have a periodic time of half a month.

Besides, we have a notable and splendid argument to
remove the scruples of those who can tolerate the revolution
of the planets round the Sun in the Copernican system, yet
are so disturbed by the motion of one Moon about the Earth,
while both accomplish an orbit of a year's length about the
Sun, that they consider that this theory of the universe must
be upset as impossible : for now we have not one planet only
revolving about another, while both traverse a vast orbit about
the Sun, but our sense of sight presents to us four satellites
circling about Jupiter, like the Moon about the Earth, while
the whole system travels a mighty orbit about the Sun in the
space of twelve years.

Galileo's discoveries and ideas were condemned by the Church, but
they could not be suppressed. Thanks to one of his telescopes, Johann
Kepler was able to make accurate additions to the extensive observations
of the positions of the planets already compiled by his master, Tycho
Brahe, and then, by a stroke of imaginative insight, to summarize their
motions in simple laws. Briefly, they moved round the sun in elliptical
paths with the sun at one focus of the ellipse ; the lines joining the planets
to the sun swept through equal areas in equal times ; and the squares
of their periods of revolution round the sun were in the same ratio as
the cubes of their distances.

The motions of the planets were thus summarized, but they were
not yet explained. In the year in which Galileo died Isaac Newton was
born, and to his genius is due the explanation which lasted until the
discovery of Relativity in the present century. *How did Newton explain
the movements of the planets round the sun ?*

According to Galileo's theory of inertia the planets would move in a
straight line unless acted on by some force. Notions of an attractive or
" centripetal " force were in the air, and, as a young man of twenty-
three, Newton himself tells us, he " began to think of gravity extending
to the orb of the moon, and having found out how to estimate the force

. . . from Kepler's Rule . . . I deduced that the forces which keep the planets in their orbs must be reciprocally as the squares of their distances from the centres about which they revolve : and thereby compared the force requisite to keep the moon in her orb with the force of gravity at the surface of the earth, and found them answer pretty well ". He returned to this subject after a long interval, and finally, twenty years after the first vital step had been taken, he went thoroughly into the theory and wrote his *Mathematical Principles of Natural Philosophy*. Extracts from this are given in answer to the question, and if they should prove rather austere we must remember that Newton wrote in Latin, as was usual in those days, and with the sole aim of convincing contemporary mathematicians of the soundness of his reasoning. Most of them found him just as difficult to follow as Einstein is to-day.

ISAAC NEWTON

THE THEORY OF GRAVITATION

THAT there are centripetal forces actually directed to the bodies of the sun, of the earth, and other planets, I thus infer.

The moon revolves about our earth, and by radii drawn to its centre describes areas nearly proportional to the times in which they are described, as is evident from its velocity compared with its apparent diameter ; for its motion is slower when its diameter is less (and therefore its distance greater), and its motion is swifter when its diameter is greater.

The revolutions of the satellites of Jupiter about that planet are more regular ; for they describe circles concentric with Jupiter by equable motions, as exactly as our senses can distinguish.

And so the satellites of Saturn are revolved about this planet with motions nearly circular and equable, scarcely disturbed by any eccentricity hitherto observed.

That Venus and Mercury are revolved about the sun, is

demonstrable from their moon-like appearances : when they shine with a full face, they are in those parts of their orbs which in respect of the earth lie beyond the sun ; when they appear half full, they are in those parts which lie over against the sun; when horned, in those parts which lie between the earth and the sun ; and sometimes they pass over the sun's disk, when directly interposed between the earth and the sun.

And Venus, with a motion almost uniform, describes an orb nearly circular and concentric with the sun.

But Mercury, with a more eccentric motion, makes remarkable approaches to the sun, and goes off again by turns ; but it is always swifter as it is near to the sun, and therefore by a radius drawn to the sun still describes areas proportional to the times.

Lastly, that the earth describes about the sun, or the sun about the earth, by a radius from the one to the other, areas exactly proportional to the times, is demonstrable from the apparent diameter of the sun compared with its apparent motion.

These are astronomical experiments ; from which it follows, by propositions 1, 2, and 3 in the first book of our *Principles*, and their corollaries, that there are centripetal forces actually directed (either accurately or without considerable error) to the centres of the earth, of Jupiter, of Saturn, and of the sun. In Mercury, Venus, Mars, and the lesser planets, where experiments are wanting, the arguments from analogy must be allowed in their place. . . .

Since the action of the centripetal force upon the bodies attracted is, at equal distances, proportional to the quantities of matter in those bodies, reason requires that it should be also proportional to the quantity of matter in the body attracting.

For all action is mutual, and (by the third law of motion) makes the bodies mutually to approach one to the other, and therefore must be the same in both bodies. It is true that we may regard one body as attracting, another as attracted ; but this distinction is more mathematical than natural. The attraction is really common of either to other, and therefore of the same kind in both.

And hence it is that the attractive force is found in both. The sun attracts Jupiter and the other planets ; Jupiter attracts its satellites ; and, for the same reason, the satellites act as well upon one another as upon Jupiter, and all the planets mutually one upon another. . . .

Perhaps it may be objected that, according to this philosophy all bodies should mutually attract one another, contrary to the evidence of experiments in terrestrial bodies ; but I answer that the experiments in terrestrial bodies come to no account ; for the attraction of homogeneous spheres near their surfaces are as their diameters. Whence a sphere of one foot in diameter, and of a like nature to the earth, would attract a small body placed near its surface with a force twenty million times less than the earth would do if placed near its surface ; but so small a force could produce no sensible effect. If two such spheres were distant but by $\frac{1}{4}$ of an inch they would not, even in spaces void of resistance, come together by the force of their mutual attraction in less than a month's time ; and less spheres will come together at a rate yet slower, viz. in the proportion of their diameters. Nay, whole mountains will not be sufficient to produce any sensible effect. A mountain of an hemispherical figure, three miles high, and six broad, will not, by its attraction, draw the pendulum two minutes out of the true perpendicular ; and it is only in the great bodies of the planets that these forces are to be perceived. . . .

As the parts of the earth mutually attract one another, so do those of all the planets. If Jupiter and its satellites were brought together, and formed into one globe, without doubt they would continue mutually to attract one another as before. And, on the other hand, if the body of Jupiter was broken into more globes, to be sure, these would no less attract one another than they do the satellites now. From these attractions it is that the bodies of the earth and all the planets effect a spherical figure, and their parts cohere, and are not dispersed through the æther. . . .

Wherefore the absolute force of every globe is as the quantity of matter which the globe contains ; but the motive force by which every globe is attracted towards another, and which, in terrestrial bodies, we commonly call their weight, is as the content under the quantities of matter in both globes applied to the square of the distance between their centres, to which force the quantity of motion, by which each globe in a given time will be carried towards the other, is proportional. And the accelerative force, by which every globe according to its quantity of matter is attracted towards another, is as the quantity of matter in that other globe applied to the square of the distance between the centres of the two ; to which force, the velocity by which the attracted globe will, in a given time, be carried towards the other is proportional. And from these principles well understood, it will now be easy to determine the motions of the celestial bodies among themselves. . . .

Thus I have given an account of the system of the planets. As to the fixed stars, the smallness of their annual parallax proves them to be removed to immense distances from the system of the planets : that this parallax is less than one minute is most certain ; and from thence it follows that the distance of the fixed stars is above 360 times greater than the distance of Saturn from the sun. Such as reckon the earth one of the planets, and the sun one of the fixed stars, may remove the fixed stars to yet greater distances by the following arguments : from the annual motion of the earth there would happen an apparent transposition of the fixed stars, one in respect of another, almost equal to their double parallax ; but the greater and nearer stars, in respect of the more remote, which are only seen by the telescope, have not hitherto been observed to have the least motion. If we should suppose that motion to be but less than 20″, the distance of the nearer fixed stars would exceed the mean distance of Saturn by above 2000 times. . . .

The fixed stars being, therefore, at such vast distances from one another, can neither attract each other sensibly, nor be attracted by our sun. But the comets must unavoidably be acted on by the circum-solar force; for as the comets were placed by astronomers above the moon, because they were found to have no diurnal parallax, so their annual parallax is a convincing proof of their descending into the regions of the planets. For all the comets which move in a direct course, according to the order of the signs, about the end of their appearance become more than ordinarily slow, or retrograde, if the earth is between them and the sun; and more than ordinarily swift if the earth is approaching to a heliocentric opposition with them. Whereas, on the other hand, those which move against the order of the signs, towards the end of their appearance, appear swifter than they ought to be if the earth is between them and the sun; and slower, and perhaps retrograde, if the earth is in the other side of its orbit. This is occasioned by the motion of the earth in different situations. If the earth go the same way with the comet, with a swifter motion, the comet becomes retrograde; if with a slower motion, the comet becomes slower however; and if the earth move the contrary way, it becomes swifter; and by collecting the differences between the slower and swifter motions, and the sums of the more swift and retrograde motions, and comparing them with the situation and motion of the earth from whence they arise, I found, by means of this parallax, that the distances of the comets at the time they cease to be visible to the naked eye are always less than the distance of Saturn, and generally even less than the distance of Jupiter.

Newton's laws of motion and his theory of universal gravitation were at first accepted only slowly, but the mathematicians qualified to judge them were quick to see their power and their beauty, and to apply them to a more detailed survey of the solar system. And a century and a half later there came a most dramatic proof of their validity in the search for, and the discovery of Neptune, a hitherto unknown planet, just in the region where Newtonian theory said it should be at that time.

The mechanism of the universe having been explained, interest shifted to allied topics, and notably to the problem of how the solar system had been formed, and how the sun managed to maintain its steady output of heat and light. Laplace, a great French mathematician, used Newton's mechanics to work out the effects of the planets on each other, and thus to provide a detailed explanation of their " perturbations " as the little irregularities in their motions are called. And on the same foundation he built up the " nebular " theory of the origin of the solar system. This was felt to be unsatisfactory, but it did suggest an answer to the question *Whence comes the heat of the sun ?* This answer is given in the famous Heidelberg Address of H. F. Helmholtz, a great German scientist who made striking discoveries in the physiology of vision and of hearing, and who, with Joule and Kelvin, established the law of the conservation of energy.

H. F. HELMHOLTZ

THE MYSTERY OF CREATION

ALL life and all motion on our earth is, with few exceptions, kept up by a single force, that of the sun's rays, which bring us light and heat. They warm the air of the hot zones ; this becomes lighter and ascends, while the colder air flows from the poles. Thus is formed the great circulation of the passage-winds. Local differences of temperature over land and sea, plains and mountains, disturb the uniformity of this great motion, and produce for us the capricious change of winds. Warm aqueous vapours ascend with the warm air, become condensed into clouds, and fall in the cooler zones, and upon the snowy tops of the mountains, as rain and as snow. The water collects in brooks, in rivers, moistens the plains, and makes life possible ; crumbles the stones, carries their fragments along, and thus works at the geological transformation of the earth's surface. It is only under the influence of the sun's rays that the variegated covering of plants of the earth grows ; and while they grow, they accumulate in their structure organic matter, which partly serves the whole animal kingdom as food, and serves man more particularly as fuel. Coals and lignites, the sources of power of our steam-engines, are remains of primitive plants, the ancient production of the sun's rays.

Need we wonder if, to our forefathers of the Aryan race in India and Persia, the sun appeared as the fittest symbol of the Deity ? They were right in regarding it as the giver of all life—as the ultimate source of almost all that has happened on earth.

But whence does the sun acquire this force ? It radiates forth a more intense light than can be attained with any terrestrial means. It yields as much heat as if 1500 pounds of coal were burned every hour upon each square foot of its surface. Of the heat which thus issues from it, the small fraction which enters our atmosphere furnishes a great mechanical force. Every steam-engine teaches us that heat can produce such force. The sun, in fact, drives on earth a kind of steam-engine whose performances are far greater than those of artificially constructed machines. The circulation of water in the atmosphere raises, as has been said, the water evaporated from the warm tropical seas to the mountain heights ; it is, as it were, a water-raising engine of the most magnificent kind, with whose power no artificial machine can be even distantly compared. I have previously explained the mechanical equivalent of heat. Calculated by that standard, the work which the sun produces by its radiation is equal to the constant exertion of 7000 horse-power for each square foot of the sun's surface.

For a long time experience had impressed on our mechanicians that a working force cannot be produced from nothing ; that it can only be taken from the stores which Nature possesses, which are strictly limited, and which cannot be increased at pleasure—whether it be taken from the rushing water or from the wind ; whether from the layers of coal, or from men and from animals, which cannot work without the consumption of food. Modern physicists have attempted to prove the universality of this experience, to show that it applies to the great whole of all natural processes, and is independent of the special interests of man. These processes have been generalized and comprehended in the all-ruling natural law of the conservation of force. No natural process, and no series of natural processes, can be found, however manifold may be the changes which take place among them, by which a motive force can be continuously produced, without a corresponding consumption. Just as the human race finds on earth but a limited supply of motive forces, capable of producing work, which it can utilize but not increase, so also must this be the case in the great whole of Nature. The universe has its definite store of force, which works in it under ever-varying forms ; is indestructible, not to be increased, everlasting and unchangeable like matter itself. It seems as if Goethe has an idea of this when he makes the earth-spirit speak of himself as the representative of natural force :—

> " In the currents of life, in the tempests of motion,
>   In the fervour of art, in the fire, in the storm,
>   Hither and thither,
>   Over and under,
>   Wend I and wander.
>   Birth and the grave,
>   Limitless ocean,
>   Where the restless wave
>   Undulates ever
>   Under and over,
>   Their seething strife
>   Heaving and weaving
>   The changes of life
>   At the whirling loom of time unawed,
>   I work the living mantle of God."

Let us return to the special question which concerns us here : Whence does the sun derive this enormous store of force which it sends out ?

On earth the processes of combustion are the most abundant source of heat. Does the sun's heat originate in a process of this kind ? To this question we can reply with a complete and decided negative, for we now know that the sun contains the terrestrial elements with which we are acquainted. Let us select from among them the two which, for the smallest mass, produce the greatest amount of heat when they combine ; let us assume that the sun consists of hydrogen and oxygen, mixed in the proportion in which they would unite to form water. The mass of the sun is known, and also the quantity of heat produced by the union of known weights of oxygen and hydrogen. Calculation shows that under the above supposition the heat resulting from their combustion would be sufficient to keep up the radiation of heat from the sun for 3021 years. That, it is true, is a long time, but even profane history teaches that the sun has lighted and warmed us for 3000 years, and geology puts it beyond doubt that this period must be extended to millions of years.

Known chemical forces are thus so completely inadequate, even on the most favourable assumption, to explain the production of heat which takes place in the sun, that we must quite drop this hypothesis.

We must seek for forces of far greater magnitude, and these we can only find in cosmical attraction. We have already seen that the comparatively small masses of shooting stars and meteorites can produce extraordinarily large amounts of heat when their cosmical velocities are arrested by our atmosphere. Now, the force which has produced these great velocities is gravitation. We know of this force as one acting on the surface of our planet when it appears as terrestrial gravity. We know that a weight raised from the earth can drive our clocks, and that in like manner the gravity of the water rushing down from the mountains works our mills.

If a weight fall from a height and strike the ground, its mass loses, indeed, the visible motion which it had as a whole —in fact, however, this motion is not lost ; it is transferred to the smallest elementary particles of the mass, and this invisible vibration of the molecules is the motion of heat. Visible motion is transformed by impact into the motion of heat.

That which holds in this respect for gravity holds also for gravitation. A heavy mass, of whatever kind, which is suspended in space separated from another heavy mass, represents a force capable of work. For both masses attract each other, and, if unrestrained by centrifugal force, they move toward each other under the influence of this attraction ; this takes place with ever-increasing velocity ; and if this velocity is finally destroyed, whether this be suddenly by collision, or gradually by the friction of movable parts, it develops the corresponding quantity of the motion of heat, the amount of which can be calculated from the equivalence previously established between heat and mechanical work.

Now we may assume with great probability that very many more meteors fall upon the sun than upon the earth, and with greater velocity, too, and therefore give more heat. Yet the hypothesis that the entire amount of the sun's heat which is continually lost by radiation is made up by the fall of meteors, a hypothesis which was propounded by Mayer, and has been favourably adopted by several other physicists, is open, according to Sir W. Thomson's investigations, to objection ; for, assuming it to hold, the mass of the sun should increase so rapidly that the consequences would have shown themselves in the accelerated motions of the planets. The entire loss of heat from the sun cannot at all events be produced in this way ; at the most a portion, which, however, may not be inconsiderable.

If, now, there is no present manifestation of force sufficient to cover the expenditure of the sun's heat, the sun must originally have had a store of heat which it gradually gives out. But whence this store ? We know that the cosmical forces alone could have produced it. And here the hypothesis, previously discussed as to the origin of the sun, comes to our aid. If the mass of the sun had been once diffused in cosmical space, and had then been condensed—that is, had fallen together under the influence of celestial gravity—if then the resultant motion had been destroyed by friction and impact with the production of heat, the new world produced by such condensation must have acquired a store of heat, not only of considerable, but even of colossal magnitude.

Calculation shows that, assuming the thermal capacity of the sun to be the same as that of water, the temperature might be raised to 28 million degrees, if this quantity of heat could ever have been present in the sun at one time. This cannot be assumed, for such an increase of temperature would offer the greatest hindrance to condensation. It is probable, rather, that a great part of this heat which was produced by condensation began to radiate into space before this condensation was complete. But the heat which the sun could have previously developed by its condensation would have been sufficient to cover its present expenditure for not less than 22 million years of the past.

And the sun is by no means so dense as it may become. Spectrum analysis demonstrates the presence of large masses of iron and of other known constituents of the rocks. The pressure which endeavours to condense the interior is about 800 times as great as that in the centre of the earth; and yet the density of the sun, owing probably to its enormous temperature, is less than a quarter of the mean density of the earth.

We may therefore assume with great probability that the sun will still continue in its condensation; even if it only attained the density of the earth—though it will probably become far denser in the interior, owing to the enormous pressure—this would develop fresh quantities of heat which would be sufficient to maintain for an additional 17 million years the same intensity of sunshine as that which is now the source of all terrestrial life.

The term of 17 million years which I have given may, perhaps, become considerably prolonged by the gradual abatement of radiation, by the new accretion of falling meteors, and by still greater condensation than that which I have assumed in that calculation. But we know of no natural process which could spare our sun the fate which has manifestly fallen upon other suns. This is a thought which we only reluctantly admit; it seems to us an insult to the beneficent Creative Power which we otherwise find at work in organisms, and especially in living ones. But we must reconcile ourselves to the thought that, however we may consider ourselves to be the centre and final object of creation, we are but as dust on the earth; which again is but a speck of dust in the immensity of space; and the previous duration of our race, even if we follow it far beyond our written history, into the era of the lake dwellings or of the mammoth, is but an instant compared with the primæval times of our planet, when living beings existed upon it whose strange and unearthly remains still gaze at us from their ancient tombs; and far more does the duration of our race sink into insignificance compared with the enormous periods during which worlds have been in process of formation, and will still continue to form when our sun is extinguished, and our earth is either solidified in cold, or is united with the ignited central body of our system.

But who knows whether the first living inhabitants of the warm sea on the young world, whom we ought perhaps to honour as our ancestors, would not have regarded our present cooler condition with as much horror as we look on a world without a sun? Considering the wonderful adaptability to the conditions of life which all organisms possess, who knows to what degree of perfection our posterity will have been developed in 17 million years, and whether our fossilized bones will not perhaps seem to them as monstrous as those of the ichthyosaurus now do; and whether they, adjusted for a more sensitive state of equilibrium, will not consider the extremes of temperature, within which we now exist, to be just as violent and destructive as those of the older geological times appear to us ? Yea, even if sun and earth should solidify and become motionless, who could say what new worlds would not be ready to develop life ? Meteoric stones sometimes contain hydrocarbons ; the light of the heads of comets exhibits a spectrum which is most like that of the electrical light in gases containing hydrogen and carbon. But carbon is the element, which is characteristic of organic compounds, from which living bodies are built up. Who knows whether these bodies, which everywhere swarm through space, do not scatter germs of life wherever there is a new world which has become capable of giving a dwelling-place to organic bodies. And this life we might perhaps consider as allied to ours in its primitive germ, however different might be the form which it would assume in adapting itself to its new dwelling-place.

However this may be, that which most arouses our moral feelings at the thought of a future, though possibly very remote, cessation of all living creation on the earth is more particularly the question whether all this life is not aimless sport, which will ultimately fall a prey to destruction by brute force. Under the light of Darwin's great thought, we begin to see that, not only pleasure and joy, but also pain, struggle, and death, are the powerful means by which Nature has built up her finer and more perfect forms of life. And we men know more particularly that in our intelligence, our civic order, and our morality we are living on the inheritance which our forefathers have gained for us, and that which we acquire in the same way will, in like manner, ennoble the life of our posterity. Thus the individual, who works for the ideal objects of humanity, even if in a modest position, and in a limited sphere of activity, may bear without fear the thought that the thread of his own consciousness will one day break. But even men of such free and large order of minds as Lessing and David Strauss could not reconcile themselves to the thought of a final destruction of the living race, and with it of all the fruits of all past generations.

As yet we know of no fact which can be established by scientific observation which would show that the finer and complex forms of vital motion could exist otherwise than in the dense material of organic life ; that it can propagate itself as the sound-movement of a string can leave its originally narrow and fixed home and diffuse itself in the air, keeping all the time its pitch, and the most delicate shade of its colour-tint ; and that, when it meets another string attuned to it, starts this again or excites a flame ready to sing to the same tone. The flame even, which of all processes in inanimate Nature is the closest type of life, may become extinct, but the heat which it produces continues to exist—indestructible, imperishable, as an invisible motion, now agitating the molecules of ponderable matter, and then radiating into boundless space as the vibration of an ether. Even there it retains the characteristic peculiarities of its origin, and it reveals its history to the inquirer who questions it by the spectroscope. United afresh, these rays may ignite a new flame, and thus, as it were, acquire a new bodily existence.

Just as the flame remains the same in appearance, and continues to exist with the same form and structure, although it draws every minute fresh combustible vapour, and fresh oxygen from the air, into the vortex of its ascending current ; and just as the wave goes on in unaltered form, and is yet being reconstructed every moment from fresh particles of water, so also in the living being it is not the definite mass of substance which now constitutes the body to which the continuance of the individual is attached. For the material of the body, like that of the flame, is subject to continuous and comparatively rapid change—a change the more rapid the livelier the activity of the organs in question. Some constituents are renewed from day to day, some from month to month, and others only after years. That which continues to exist as a particular individual is like the flame and the wave—only the form of motion which continually attracts fresh matter into its vortex and expels the old. The observer with a deaf ear only recognizes the vibration of sound as long as it is visible and can be felt, bound up with heavy matter. Are our senses, in reference to life, like the deaf ear in this respect ?

It is probable that Helmholtz's eloquent peroration will be remembered long after the main body of his address is forgotten, for his explanation of the source of the sun's energy, while true enough so far as it goes, does not go nearly far enough. It gives a supply of heat sufficient for 20 million years, but geologists and astronomers—and biologists—require a much longer period of time. The true solution to this problem only became available with the discovery of atomic energy.

You will remember that Newton mentioned the immense distances of the stars, but even Newton would be amazed at the modern revelations of the tremendous size of the universe. *How big is the universe ?* Some idea of the immensity, and of the philosophical implications following thereon, may be gathered from the following address by Dr. Hector MacPherson.

HECTOR MACPHERSON

THE UNIVERSE AS REVEALED BY MODERN ASTRONOMY

THE modern science of astronomy is little more than three centuries old. It is true that over 380 years have elapsed since the corner-stone of the edifice was laid by the publication of the great work of Copernicus, *De Revolutionibus*. In that book the quiet recluse of Frauenburg formulated the theory, which was to have so profound an influence on human thought, that the apparently immovable earth was in motion round the apparently mobile sun. But for the best part of a century the new heliocentric theory was laughed out of court as the dream of a crank, not only by the theologians, Roman and Protestant alike, but also by the professional scientists of the times, and it was only when Galileo and Kepler, working along independent lines, were driven to accept the Copernican system, and as a result of their own scientific work became its protagonists, that the heliocentric system was seriously considered. Indeed, it can only be said to have been finally adopted by the scientific world as a whole as a result of the work of Newton.

It is necessary for us to remember, therefore, that at the time when the creeds of the Reformation and the Counter-Reformation—those Protestant and Romanist Confessions which are still the standards of organized Christianity to-day—were formulated, the geocentric system was unquestioningly accepted. The earth was regarded as the only world, the end and aim of creation round which all the heavenly bodies revolved, and for the benefit of the inhabitants of which the sun had been created to illuminate the day and the moon and stars to mitigate the darkness of the night. To this view the Reformers clung as tenaciously as the Inquisitors. Luther, in his usual pugnacious style, referred to Copernicus as a fool who dared to contradict the Bible, an " upstart astrologer " who " set his own authority above that of sacred Scripture ". Melanchthon gravely deplored the lack of decency of those who set forth such theories ; while Calvin clinched the matter, as he thought, by putting the scathing question—" Who will venture to place the authority of Copernicus above that of the Holy Spirit ? "

These efforts to stem the current of human progress were about as efficacious as the legendary attempt of King Canute to stay the flowing tide. During the past three centuries the history of astronomy has been the record of the continuous extension of our knowledge of the universe both in space and time, and the cosmos which confronts the astronomer to-day is inconceivably vaster than the tiny world familiar to the ancients and regarding which the learned men even up to the seventeenth century believed themselves to know so much. Certainly the outstanding fact which the study of astronomy drives home upon our minds with irresistible force—the truth which strikes even the juvenile inquirer as he reads for the first time such a book as Ball's *Star-Land*—is the vastness of the visible universe. The earth, which our forefathers believed to be the centre of all things, is shown by modern astronomy to be but one planet among others revolving round the sun ; and when we realize the magnitude of the sun and of the greater planets, and the vastness of the distances separating the various worlds from the sun and from one another, we understand how insignificant a rôle our world plays even in the solar system. The earth is the centre, not of the universe, but of a little sub-system within the greater solar system—the Terrestrial or Earth-Moon system. Only in comparison with the moon does the earth appear large and important ; and our satellite is the one astronomical body which is comparatively close to us. Only 238,000 miles—a mere trifle so far as celestial distances go—separate us from that mass of matter which appears to be after all, to quote the picturesque phrase of Flammarion, nothing but a " detached continent ". The small domain over which our earth holds sway is less than 500,000 miles in diameter, and when we are dealing with thousands of miles we are still within the realm of the conceivable.

When we come to consider the magnitude and distance of the sun we find ourselves projected into a new world of size and distance altogether. In volume this mighty globe exceeds our earth, 1,300,000 times ; and while we measure the distance of the moon in thousands of miles, our unit for the inter-planetary distances is the million. The mean distance of the earth from the sun, as every schoolboy knows, is 93 million miles. But this is one of the smaller distances in the solar system, for the earth is comparatively close to the sun. Neptune, the outermost known planet, revolves at a distance of 2700 million miles, so that the sun's domain—leaving out of consideration the orbits of the erratic and unstable cometary bodies—is over 5000 million miles in diameter.[1]

[1] In March 1930 a new planet, Pluto, was discovered. It is about 3700 million miles from the sun, and takes about 250 years to move once round the sun.

The eight primary planets fall into two distinct groups. They have been designated as the outer and inner planets, and this is a convenient classification ; but they may be grouped also according to their size and physical condition, and following up the recent classification of the stars, I have suggested that the two groups should be designated the giant and dwarf planets respectively. Into which of these groups does our earth fall ? Not into the giant group. Jupiter, Saturn, Uranus, Neptune vastly exceed our world in size—the volume of Jupiter being 1300 times that of the earth. The only position of dignity which our world may be said to occupy is the primacy among the dwarf planets, being slightly larger than Venus, considerably larger than Mars, and much larger than Mercury. The earth, then, may be defined as a dwarf planet revolving round the sun.

The solar system, vast though it is, is not co-extensive with the universe. The astronomers of the eighteenth century, preoccupied though they were with the motions of the planets and the verification of the Newtonian law, realized this in a dim, vague kind of way. But it is to the elder Herschel that the world owes the pioneer work which has issued in our present-day cosmology. His discovery of the motion through space of the sun, carrying with it the earth and all the planets and their satellites, indicated the essential kinship of the sun and the so-called " fixed stars ", some of which were, even in Herschel's time, known to be in quite rapid motion. And within the past sixty years, the spectroscope has proved beyond a doubt that there is nothing unique about the sun. It is simply a star, one among a vast number—75 million at a minimum estimate, and 1000 million at a maximum. Its importance to us consists simply in the fact that our earth is one of its small satellites. Intrinsically it is not important.

One of the broad generalizations of modern astronomy is that associated with the name of Professor H. N. Russell, of Princeton. Professor Russell finds that the great mass of the

stars fall into two well-defined groups, which he designates as
giants and dwarfs. While the giants—which form the majority
of the brilliant first-magnitude stars which are familiar even to
the casual star-gazer, such as Antares, Betelgeux, Aldebaran,
Arcturus, and Rigel—are tens and hundreds of millions of
miles in diameter, the diameters of the dwarfs are to be
measured in hundreds of thousands of miles. Not quite four
years ago, the astronomers at the Mount Wilson Observatory,
in California, succeeded in measuring the diameter of the bright
star Betelgeux, which they found to be 273 million miles.
Some time later they found that Antares is a still mightier star,
its diameter being no less than 430 million miles. And just
the other day Dr. Shapley, the director of Harvard College
Observatory, announced that in all likelihood the giant stars
in the Nubecula Minor are still larger, some of them being
about 1000 million miles in diameter. These amazing facts
show us that the sun, vast though it is relatively to our little
world, falls into the group of dwarf stars. It is a large dwarf,
certainly, but a dwarf nevertheless. So our earth's status in
the stellar system proves to be that of a dwarf planet revolving
round a dwarf star.

The vastness of the material universe is perhaps still more
forcibly impressed on us by the distances which separate the
stars from one another. The nearest stars, the bright first-
magnitude double star, Alpha Centauri, and the tiny dwarf
Proxima, which seems to be physically connected with it, are
distant 25 billions of miles. So distant are these stars, indeed,
that light, which travels from the moon to the earth in a second
and a half, and from the sun in eight minutes, requires four
years for the journey. And these Centaurus stars are our
nearest neighbours. The familiar brilliants of our evening sky
are situated at much greater distances. The nearest of the
giants would appear to be Vega, about 30 light-years away—a
light-year is about 6 billion miles—while the super-giants,
Antares and Betelgeux, are distant over 100 light-years away.
At still greater distances are the familiar cluster of the Pleiades,

and the bright groups which we know as Ursa Major and Orion. These appear to be at least 600 light-years away.

The researches of Professor Shapley have resulted in a still further extension of our knowledge of stellar distances. Astronomers at the beginning of the century estimated the diameter of the entire stellar system at about 10,000 light-years at the outside. But Dr. Shapley some years ago discovered in the dense star-clouds of the Galaxy faint blue stars, evidently giants, which cannot be nearer than 15,000 light-years. 60,000 light-years away from the sun is the centre of gravity of the stellar system, while the whole great assemblage of stars has the shape of a flattened disc, about 6000 light-years in thickness and about 300,000 in diameter—100,000 times greater than was supposed at the beginning of the century. Outside of this main sidereal system there are subordinate or satellite systems —" island universes ", using that designation in a literal sense. By mutually confirmatory methods, Dr. Shapley has fixed the distances of eighty-six of these globular clusters. The nearest is 22,000 light-years away, and the most distant 220,000. Not so authoritative, perhaps, is an estimate made by the same astronomer of the distance of a still more distant outlying cluster, which he has computed to be 1 million light-years, or 6 million billion miles.

The vexed question of the finitude or infinity of the physical universe, or rather the question as to whether it is limited or unlimited, remains unanswered. What we call the stellar system is most certainly finite, but whether or not it is but one island amid myriads of others we do not know. The spiral nebulæ, which about ten or twelve years ago were thought by many astronomers to be external galaxies, are most probably not so, but truly nebulous masses. If certain deductions from the Einstein theory be reliable, space—at least that space with which we are familiar—is finite. Einstein has indeed fixed an upper limit—a maximum extent of the space-time continuum, 1000 million light-years in circumference. But even if this is so, if the space we know in association with matter is " finite yet unbounded ", we have no reason to believe that it exhausts the All. It may be but a temporary manifestation. There may be more than poetry in Shelley's fine words :—

" What is heaven ?   A globe of dew,
A frail and fading sphere,
With ten millions gathered there
To tremble, gleam, and disappear."

And if we are confronted by virtual illimitability in space, we are faced no less with virtual limitlessness in time. Religious people even a hundred years ago were accustomed to think of the world as 6000 years old. Archbishop Ussher fixed the date of the creation of the world out of nothing, declaring for about 4000 years B.C. Dr. John Lightfoot, Vice-Chancellor of Cambridge University, one of the most erudite seventeenth-century scholars, solemnly calculated the date of creation as the month of October 4004—man being created by the Trinity on the twenty-third of that month at nine o'clock in the morning. By the mid-nineteenth century, Lord Kelvin fixed a maximum of 20 million years for the life of the sun, with a shorter period for that of our world. His estimates, however, have been rendered obsolete by investigations in the realm of stellar evolution and by the recognition by physicists of the hitherto disregarded factor of inter-atomic energy. Dr. Shapley's study of the physical condition of the stars in the nearest cluster, 22,000 light-years away, and the most distant system separated by 220,000 light-years, indicates that the stars in the one cluster appear to be at the same stage of development as the stars in the other, though the light from the one cluster has been so much longer on its way than the light from the other. 200,000 years, then, is a negligible quantity in the life of a star —a mere tick of the clock. Hundreds and thousands of millions of years have been required for the evolution of our own world. Possibly a billion years is the unit of time in the life-history of a star. Finite it may be, nevertheless the visible universe presents to the mind of man an impression of incomprehensible

vastness alike in space and time. " The spirit of man acheth
with this infinity ", as Richter truly said. Flammarion does
not over-estimate the truth when he says :—

> " Such is the aspect, grand, splendid, and sublime, of
> the universe which flies through space before the dazzled
> and stupefied gaze of the terrestrial astronomer, born to-day
> to die to-morrow on a globule, lost in the infinite night."

It must be confessed that the first effect of this impression
of vastness must be one of unsettlement. Little wonder that
conventional theologians fought this new cosmology long and
fiercely. The new cosmology does not, as they thought,
contradict the Confessions and the Creeds, but it somehow
renders them meaningless ; for the Power which the cosmology
and cosmogony of to-day alike hint at is as much greater than
the God of the Council of Trent or the Westminster Assembly
as the universe of to-day is wider than the universe of Luther.
When Shelley said of the God of the Jews, or rather he might
have said the God of the Oxford theologians of the early
nineteenth century, that " the works of His fingers have borne
witness against Him ", he was not blaspheming, however
unfortunate the tone of those immature notes to " Queen
Mab ". He was simply stating the actual conclusion to be
derived from the impact of modern astronomy on scholastic
theology, and particularly on crude theories of the Atonement.
I do not know of a better statement of the bearing of modern
astronomical concepts on our theological outlook than the words
of Dr. Beard in his Hibbert Lecture on " The Reformation in
relation to Modern Thought and Knowledge " :—

> " How shall we rise to the thought of Him who is the
> Lord of innumerable worlds, the Ruler of the boundless
> spaces, the Master of the Eternal years ? Did then God,
> and such a God as the all of things prove He must be, die
> for us ? I say it with the deepest respect for the religious
> feelings of others, but I cannot but think that the whole

system of Atonement, of which Anselm is the author, shrivels into inanity amid the light, the space, the silence of the stellar worlds."

And what Dr. Beard says of Atonement is true of much of our theological thinking on miracle, Christology, eschatology. While it is true that scientific and religious knowledge are in the last analysis incommensurable, it is also true that he who has in any real sense grasped the import of modern cosmology will hesitate before he maintains that in the subsidiary dogmas of theology we have absolute truth or that they are more than imperfect formulations, relative to our experience.

Is it accurate, then, to say that the foundations of faith are cut away from us in " the height, the depth, the gloom, the glory " ? Does our spark of being wholly vanish in the depths and heights ? By no means. Our cursory sketch of the visible universe must have left on our minds certain impressions.

1. The most definite impression is surely that this stupendous system of stars and worlds is not self-explanatory. As a prominent American astronomer said about a year ago :—

" My own deliberately formed judgment, purely personal, not susceptible of proof, perhaps, is that so tremendous a cosmos must have divinity in it or over it ; my reason rebels at the assumption that it is purely materialistic, the result of the chance concentration of self-created physical forces."

The old cosmological proof of the divine existence is no longer valid, it is true, as a logical proof. At the same time the universe as a whole does strongly suggest that it is not self-explanatory.

" The cosmological argument," as has been well said by Professor Upton, " when it takes the shape of asserting that a unitary ground and cause is needed to account for and render intelligible this entire infinite series of dynamic activities and phenomenal changes which constitute the universe rests still, I believe, upon a solid foundation of logical necessity ".

or as another philosopher has said with equal truth :—

" The earliest assumption of human thought, that an adequate producing power is implied in the existence of what we see, is also the testimony of the visible universe, with its immeasurable vastness and its infinite variety."

And even as the universe is immeasurable, its fundamental cause is immeasurable. The universe is a manifestation of immeasurable power—of that " Infinite and Eternal Energy " before which even the professedly agnostic Spencer stood in awe and reverence. The primitive emotions of wonder and worship which the star-lit sky aroused in our ancestors it still arouses, only the wonder and worship are heightened and deepened by the assured results of modern astronomy.

2. But astronomy reveals to us other aspects of the universe besides vastness, and from these aspects we have something to learn. Without making any illegitimate assumptions, we can glean something further from the starry heavens concerning the Power beyond.

Modern astronomy has demonstrated the oneness of the universe. When Newton formulated his law of gravitation he showed that the sun and the planets formed one system, that the earth and its fellow-worlds were members of one family, performing their revolutions in obedience to one law. And when, in 1802, Herschel discovered the binary stars, it was further proved that the stellar system was one family, and that throughout the whole stellar system masses of matter obeyed a uniform law. Further, the modern branch of astronomy known as astrophysics has still further emphasized the kinship between the earth and the sun on the one hand and the stars on the other. The dark lines in the solar spectrum tell unmistakably of the existence in the solar atmosphere, in gaseous form, of the very elements with which we are familiar here upon earth—sodium and potassium, calcium, iron, hydrogen.

The day-star is akin to us, bone of our bone and flesh of our flesh. And while it is profoundly true that " one star differeth from another star in glory ", the same chemical elements are yet to be found in the atmospheres of giant and dwarf stars alike. These elements exist in different proportions and under varied conditions of temperature and pressure, but their presence is clearly demonstrated by the faint lines in the stellar spectra. Matter, then, is subject to the same laws in the most distant parts of the universe as here on earth. The same laws hold sway and the same processes are going on. The new physics does not speak of gravitation as a force, but as a property of space, but whatever gravitation may be on the ultimate analysis, it is cosmos-wide in its scope. What we look out on from our vantage-point on earth is a unified universe—one in law, one in substance, one in process. Given the same condition, the same results will follow. And in passing it may be observed we cannot but be struck by the absurdity of believing that only on one small speck of matter has the cosmic process resulted in what we know as life and intelligence. Apart from the interpretation of the mysterious markings on our neighbour world Mars which the keen eyes of Schiaparelli and Lowell discerned and which, it may be, seems to hint at the existence of cosmical cousins out there in space, the plurality of worlds would appear to be the necessary corollary of the oneness of the universe. Even if, as some astronomers think, there are no planets revolving round the double stars, even if the solar system had an unusual origin, even if only one star in a thousand can ever be the centre of a system of planets—all of which are highly controversial propositions—we are faced with the likelihood that life has appeared in many different regions of the stellar system. The bearing of this on the dogmas of orthodox theology is obvious.

The universe, then, is a universe and not a multiverse, and if we are permitted to say that the system of the stars speaks to us of Power immeasurable, we can with equal legitimacy infer that it speaks of one Power. It is at theism and not at polytheism, at one Creator, not at many, that the universe hints; and certainly not at a limited or finite God.

3. Further, the universe is quite evidently an ordered universe. A cursory glance, it is true, indicates chaos and confusion; but with the progress of science, this chaos and confusion gives place to order and harmony. For instance, the seemingly hopeless tangle of the planetary motions was resolved into the simplest of systems by the formulation of the Copernican theory. The apparently complex motions of the planets are the sum of the motions of the individual planets plus that of the earth; and the actual movements are the necessary consequences of the Keplerian and the Newtonian laws. The motions of the stars are much more complicated, but since the time of Herschel it has been evident that the stellar system will one day yield up the secret of its structure. Just as in the solar system, so in the greater stellar system the stars are concentrated to the plane of the Galaxy or Milky Way. And in recent years much progress has been made in unravelling the apparently tangled skein of the stellar motions. One of the greatest discoveries of the century has been the detection, chiefly through the labours of Kapteyn, Eddington, and Dyson, of the fact that the stars in the vicinity of the sun belong to one or other of two great streams of stars moving in opposite directions—a fact which seems to find an explanation in the tentative cosmology of Dr. Shapley, who views it as the effect of the mingling of two clusters of stars, to one of which our sun belongs. It may take many decades or centuries before the construction of the universe is even approximately known, but the problem is obviously not insoluble.

What, then, does the fact that the universe is an ordered universe teach us? That ultimately the universe is understandable. And if we are entitled to say that the universe speaks to us of Power, and of one Power, we are no less justified in saying that the Power hinted at is not blind Power, but a Power based on reason, whose thoughts, in Kepler's fine phrase, we think afterwards in the process of discovery.

4. We are, I think, entitled to go a step further. Astronomical science tells us a little more. When the average man thinks of evolution, he thinks of biology and Darwinism and Mendelism, of the struggle for existence and the origin of species. But, truth to tell, the first hint of evolution which the mind of man received was not from the earth beneath, but from the heaven above. In the closing years of the eighteenth century the idea dawned on the mind of Herschel that some at least of the faint filmy nebulæ which his great reflectors revealed in such lavish profusion were masses of primeval world-stuff, from which the stars were in the course of ages evolved. And Laplace, contemplating the finished article before him, namely the solar system, reasoned backwards to a time when the sun was a vastly extended nebula, and when this world was a " fluid haze of light ". After many vicissitudes the idea of the development of stars out of nebulæ was finally accepted in the last half of the nineteenth century and is to-day established on unassailable foundations. That the primeval world-stuff is believed to be exemplified in the dark nebulous matter of which the luminous nebulæ are but exceptional specimens does not invalidate the nebular hypothesis in the least.

The classification of stars according to their physical condition, which the invention of the spectroscope rendered possible, gave the clue to Vogel in 1874 that a spectral classification might be taken to represent the order of stellar evolution. And while particular systems of classifications have come and gone, there can be no doubt that the vast majority of the stars in the sky can be arranged spectroscopically in an evolutionary series. The theory of Professor Russell divides the stars into the two classes of giants and dwarfs—the former being great gaseous masses in process of contraction and still growing hotter while the latter have passed the meridian of stellar life, and are still contracting but growing cooler, as our sun is doing. But the mere fact that the stars can be arranged in an evolutionary sequence from the simple to the complex, from the shapeless cloud of cosmical dust to the steady star, fitted to be the centre of a system of worlds, strongly suggests purpose, cosmos-wide in its scope. What Professor Arthur Thomson says of the organic realm is equally true of the inorganic :—

" Only a system with order and progress in the heart of it could elaborate itself so perfectly and so intricately. There is assuredly much to incline us to ' assert Eternal Providence and justify the ways of God to men '."

If the universe hints at the existence of a great causal Power, one Power, one understandable Power, it hints as strongly that this Power is working with purpose according to plan.

I venture to say that while there is not, and in the nature of the case cannot be, any logical proof of the existence of God, the universe revealed by modern astronomy gives us this impression of Power, one Power, one wise Power, one wise purposive Power behind all phenomena. I believe that the facts of astronomical science are not only not hostile to Christian theism, but that the queen of the sciences is in a very real sense the handmaid of faith. The God of the scientist, to quote a notable sentence of John Fiske, the disciple of Spencer, " is still and must ever be the God of the Christian, though free from the illegitimate formulas by the aid of which theology has sought to render Deity comprehensible ".

Theism—Christian theism—has nothing to fear from astronomical science, but rather much to gain by the assimilation of the assured results of astronomy. While a little knowledge may be conducive to atheism, as Bacon truly observed, " much natural philosophy and wading deep into it will bring about men's minds to religion ".

Dr. MacPherson has introduced a strongly philosophical note into our reading, and while our minds are tuned to high endeavour we might try to learn something about relativity.

Nineteenth-century scientists had decided that light was a wave motion in what they called the " ether". This ether pervaded all space, since light reached us from the most distant stars. The earth, therefore, in its motion round the sun must travel through the ether, and a ray of light travelling to a mirror and back again should have a slightly different velocity according to whether it travelled up and down the ether stream or across it. Two American scientists, Michelson and Morley, devised apparatus sufficiently sensitive to measure the expected difference, but much to their surprise no such difference could be detected. The velocity of light was always the same, always 186,000 miles per second. Various partial explanations were offered, but in 1905 Einstein gave us his special theory of relativity which suggested that the space and time which we measure in such experiments depend on the motion of the observer. Three years later Minkowski showed that we could only obtain a true picture of the universe by giving up our habit of separating space and time, and by considering a four-dimensional continuum. Just as we include both space and time in our conception of speed, so that we cannot have any idea of speed without uniting distance and time, so we should try to visualize the universe as an equally close union of length, breadth, height, and time. In 1915 Einstein gave us the General Theory of Relativity which brought gravitation into the scope of the new ideas. Since the effects of gravity may be exactly duplicated by an acceleration, as in a lift, or a change of path, as in a swooping aeroplane, it is possible that a curvature of space-time may be the real cause of " gravitation ".

How did Relativity modify Newton's system? The answer is given, very briefly, in the following extracts from Sir Arthur Eddington's *Space, Time, and Gravitation,* and in them, too, we shall see how careful observations with delicate instruments provided proof, first from the deflection of a ray of light as it passed near the sun, and secondly from the changes in the perihelion (closest approach to the sun) in the orbit of Mercury, that Einstein was right.

SIR ARTHUR EDDINGTON

GRAVITATION AND RELATIVITY

WAS there any reason to feel dissatisfied with Newton's law of gravitation ?

Observationally it had been subjected to the most stringent tests, and had come to be regarded as the perfect model of an exact law of nature. The cases where a possible failure could be alleged were almost insignificant. There are certain unexplained irregularities in the moon's motion; but astronomers generally looked—and must still look—in other directions for the cause of these discrepancies. One failure only had led to a serious questioning of the law; this was the discordance of motion of the perihelion of Mercury. How small was this discrepancy may be judged from the fact that, to meet it, it was proposed to amend *square* of the distance to the 2·00000016 power of the distance. Further it seemed possible, though unlikely, that the matter causing the zodiacal light might be of sufficient mass to be responsible for this effect.

The most serious objection against the Newtonian law as an exact law was that it had become ambiguous. The law refers to the product of the masses of the two bodies; but the mass depends on the velocity—a fact unknown in Newton's day. Are we to take the variable mass, or the mass reduced to rest? Perhaps a learned judge, interpreting Newton's statement like a last will and testament, could give a decision; but that is scarcely the way to settle an important point in scientific theory.

Further, *distance*, also referred to in the law, is something relative to an observer. Are we to take the observer travelling with the sun or with the other body concerned, or at rest in the æther or in some gravitational medium? . . .

It is often urged that Newton's law of gravitation is much simpler than Einstein's new law. That depends on the point of view; and from the point of view of the four-dimensional world Newton's law is far more complicated. Moreover, it will be seen that if the ambiguities are to be cleared up, the statement of Newton's law must be greatly expanded.

Some attempts have been made to expand Newton's law on the basis of the restricted principle of relativity alone. This was insufficient to determine a definite amendment. Using the principle of equivalence, or relativity of force, we have arrived at a definite law proposed in the last chapter. Probably the question has arisen in the reader's mind, why should it be called the law of gravitation? It may be plausible as a law of nature; but what has the degree of curvature of space-time to do with attractive forces, whether real or apparent?

A race of flat-fish once lived in an ocean in which there were only two dimensions. It was noticed that in general fishes swam in straight lines, unless there was something obviously interfering with their free courses. This seemed a very natural behaviour. But there was a certain region where all the fish seemed to be bewitched; some passed through the region but changed the direction of their swim, others swam round and round indefinitely. One fish invented a theory of vortices, and said that there were whirlpools in that region which carried everything round in curves. By-and-by a far better theory was proposed; it was said that the fishes were all attracted towards a particularly large fish—a sun-fish—which was lying asleep in the middle of the region; and that was what caused the deviation of their paths. The theory might not have sounded particularly plausible at first; but it was confirmed with marvellous exactitude by all kinds of experimental tests. All fish were found to possess this attractive power in proportion to their sizes; the law of attraction was extremely simple, and yet it was found to explain all the motions with an accuracy never approached before in any scientific investigations. Some fish grumbled that they did not see how there could be such an influence at a distance; but it was generally agreed that the influence was communicated through the ocean and might be better understood when more was known about the nature of water. Accordingly, nearly every fish who wanted to explain the attraction started by proposing some kind of mechanism for transmitting it through the water.

But there was one fish who thought of quite another plan. He was impressed by the fact that whether the fish were big or little they always took the same course, although it would naturally take a bigger force to deflect the bigger fish. He therefore concentrated on the courses rather than on the forces.

And then he arrived at a striking explanation of the whole thing. There was a mound in the world round about where the sun-fish lay. Flat-fish could not appreciate it directly because they were two-dimensional ; but whenever a fish went swimming over the slopes of the mound, although he did his best to swim straight on, he got turned round a bit. (If a traveller goes over the left slope of a mountain, he must consciously keep bearing away to the left if he wishes to keep to his original direction relative to the points of the compass.) This was the secret of the mysterious attractions, or bending of the paths, which was experienced in the region.

The parable is not perfect, because it refers to a hummock in space alone, whereas we have to deal with hummocks in space-time. But it illustrates how a curvature of the world we live in may give an illusion of attractive force, and indeed can only be discovered through some such effect. . . .

A ray of light passing near a heavy particle will be bent, firstly, owing to the non-Euclidean character of the combination of time with space. This bending is equivalent to that due to Newtonian gravitation, and may be calculated in the ordinary way on the assumption that light has weight like a material body. Secondly, it will be bent owing to the non-Euclidean character of space alone, and this curvature is additional to that predicted by Newton's law. If then we can observe the amount of curvature of a ray of light, we can make a crucial test of whether Einstein's or Newton's theory is obeyed. . . .

The bending affects stars seen near the sun, and accordingly the only chance of making the observation is during a total eclipse when the moon cuts off the dazzling light. Even then there is a great deal of light from the sun's corona which stretches far above the disc. It is thus necessary to have rather bright stars near the sun, which will not be lost in the glare of the corona. Further the displacements of these stars can only be measured relatively to other stars, preferably more distant from the sun and less displaced ; we need therefore a reasonable number of outer bright stars to serve as reference points.

In a superstitious age a natural philosopher wishing to perform an important experiment would consult an astrologer to ascertain an auspicious moment for the trial. With better reason, an astronomer to-day consulting the stars would announce that the most favourable day of the year for weighing light is May 29. The reason is that the sun in its annual journey round the ecliptic goes through fields of stars of varying richness, but on May 29 it is in the midst of a quite exceptional patch of bright stars—part of the Hyades—by far the best star-field encountered. Now if this problem had been put forward at some other period of history, it might have been necessary to wait some thousands of years for a total eclipse of the sun to happen on the lucky date. But by strange good fortune an eclipse did happen on May 29, 1919. . . .

Attention was called to this remarkable opportunity by the Astronomer Royal in March 1917; and preparations were begun by a Committee of the Royal Society and Royal Astronomical Society for making the observations. Two expeditions were sent to different places on the line of totality to minimize the risk of failure by bad weather. Dr. A. C. D. Crommelin and Mr. C. Davidson went to Sobral in North Brazil; Mr. E. T. Cottingham and the writer went to the Isle of Principe in the Gulf of Guinea, West Africa. . . .

It will be remembered that Einstein's theory predicts a deflection of $1''\cdot74$ at the edge of the sun, the amount falling off inversely as the distance from the sun's centre. The simple Newtonian deflection is half this, $0''\cdot87$. The final results (reduced to the edge of the sun) obtained at Sobral and Principe with their " probable accidental errors " were

$$\text{Sobral} \quad 1''\cdot98 \pm 0''\cdot12$$
$$\text{Principe} \quad 1''\cdot61 \pm 0''\cdot30$$

It is usual to allow a margin of safety of about twice the probable error on either side of the mean. The evidence of the Principe plates is thus just about sufficient to rule out the possibility of the " half deflection ", and the Sobral plates exclude it with practical certainty. The value of the material found at Principe cannot be put higher than about one-sixth of that at Sobral ; but it certainly makes it less easy to bring criticism against this confirmation of Einstein's theory seeing that it was obtained independently with two different instruments at different places and with different kinds of checks. . . .

We have seen that the swift-moving light-waves possess great advantages as a means of exploring the non-Euclidean property of space. But there is an old fable about the hare and the tortoise. The slow-moving planets have qualities which must not be overlooked. The light-wave traverses the region in a few minutes and makes its report ; the planet plods on and on for centuries, going over the same ground again and again. Each time it goes round it reveals a little about the space, and the knowledge slowly accumulates.

According to Newton's law a planet moves round the sun in an ellipse, and if there are no other planets disturbing it, the ellipse remains the same for ever. According to Einstein's law the path is very nearly an ellipse, but it does not quite close up ; and in the next revolution the path has advanced slightly in the same direction as that in which the planet was moving. The orbit is thus an ellipse which very slowly revolves.

The exact prediction of Einstein's law is that in one revolution of the planet the orbit will advance through a fraction of a revolution equal to $3v^2/C^2$, where v is the speed of the planet and C the speed of light. The earth has $1/10,000$ of the speed of light ; thus in one revolution (one year) the point where the earth is at greatest distance from the sun will move on $3/100,000,000$ of a revolution, or $0''\cdot038$. We could not detect this difference in a year, but we can let it add up for a century at least. It would then be observable but for one thing—the earth's orbit is very blunt, very nearly circular, and so we cannot tell accurately enough which way it is pointing and how its sharpest axes move. We can choose a planet with higher speed so that the effect is increased, not only because v^2 is increased, but because the revolutions take less time ; but what is perhaps more important, we need a planet with a sharp elliptical orbit, so that it is easy to observe how its apses move round. Both these conditions are fulfilled in the case of Mercury. It is the fastest of the planets, and the predicted advance of the orbit amounts to 43″ per century ; further, the eccentricity of its orbit is far greater than of any of the other seven planets.

Now an unexplained advance of the orbit of Mercury had long been known. It had occupied the attention of Le Verrier, who, having successfully predicted the planet Neptune from the disturbances of Uranus, thought that the anomalous motion of Mercury might be due to an interior planet, which was called Vulcan in anticipation. But though thoroughly sought for, Vulcan has never turned up. Shortly before Einstein arrived at his law of gravitation, the accepted figures were as follows. The actual observed advance of the orbit was 574″ per century, the calculated perturbations produced by all the known planets amounted to 532″ per century. The excess of 42″ per century remained to be explained. Although the amount could scarcely be relied on to a second of arc, it was at least thirty times as great as the probable accidental error.

The big discrepancy from the Newtonian gravitational theory is thus in agreement with Einstein's prediction of an advance of 43″ per century. . . .

The theory of relativity has passed in review the whole subject-matter of physics. It has unified the great laws, which by the precision of their formulation and the exactness of their application have won the proud place in human knowledge which physical science holds to-day. And yet, in regard to the nature of things, this knowledge is only an empty shell—a form of symbols. It is knowledge of structural form, and not knowledge of content. All through the physical world runs that unknown content, which must surely be the stuff of our consciousness. Here is a hint of aspects deep within the world of physics, and yet unattainable by the methods of physics. And, moreover, we have found that where science has progressed the farthest, the mind has but regained from nature that which the mind has put into nature.

We have found a strange foot-print on the shores of the unknown. We have devised profound theories, one after another, to account for its origin. At last, we have succeeded in reconstructing the creature that made the footprint. And Lo ! it is our own.

Einstein's earlier Special Theory of relativity gave us a new geometry of space-time ; his later General Theory showed how what we had previously called gravitational " force " could be fitted into this geometry as a characteristic of the four-dimensional continuum. And ever since he completed his work on the General Theory he has been trying to bring electro-magnetic " force " into the same fold. He has been searching for a Unified Field Theory which will describe the properties and motions of atomic particles as well as of planets and stars.

Many years ago Clerk Maxwell gave six equations which together tell us how an electro-magnetic field varies in space and time. Einstein's General Theory gave us ten functions which together characterize space-time and gravitation. His new theory gives us sixteen functions which sum up in abstruse mathematical language the geometry of the universe and the fundamental laws of physics. The Greek scientists, you may remember, were always seeking for some general system or theory which would account for all the different phenomena of nature. They failed to find it, which is not at all surprising when we know how complicated nature is. *Has Einstein realized the great ambition of the ancient Greek philosophers ?* The following lecture suggests that it is very probable.

J. BRONOWSKI

DR. EINSTEIN SUMS UP

NEARLY 300 years ago the Great Plague was raging in London and sending people into hiding up and down the country. Among those who took flight at the plague was the University of Cambridge, which packed its students off home for the best part of two years. This was how it happened that some time in 1666 a young man of twenty-three was sitting in the garden of his mother's farm in Lincolnshire, thinking about the Universe by himself, with no help from his professors. And while his thoughts were on such things as the Moon, an apple fell to the ground. Tradition has it that the apple was a Flower of Kent.

You have heard this story before—and it is a true story— about Isaac Newton and the apple. It is the most famous apple since the Garden of Eden. And it deserves its fame. In that moment of insight in the hushed orchard we are present at the birth of a great theory which now, after 300 years, is crowned by Einstein's works. But there is something even deeper. We look for an instant into the nature of science itself. Men by the thousand had thought about the Moon as Newton did, and have seen apples drop; they were just two different things that happen. Newton's inspiration was to ask himself whether these really were two quite different kinds of happening. The apple was drawn to the ground by a force; might not this force of gravity also reach far out into space? Might not this be the pull which holds the Moon in its orbit round the Earth, and the Earth in its orbit round the Sun like a stone swung in a sling?

So, suddenly, the Universe became orderly. It was seen to be held together by the forces between every two pieces of matter in it, and—this is the point—these forces were all of the same kind. Unity had been given to the tangle of events by that insight which put side by side gravity on the Earth and the path of the Moon, and saw them to be expressions of one underlying principle. That is the true nature of science: to take two kinds of happening between which there seems to be no link, and by an act of imagination show them to be aspects of one wider law. I say " an act of imagination ", which is not how we commonly speak of scientific thinking. Yet this is the essence of all such insight, which sees the deeper likeness under

what seems unlike ; and this not only in science. This gives
the shock of pleasure which we feel at the images of poetry.
When Shakespeare says of Cleopatra :—

> " The barge she sat in, like a burnish'd Throne
> Burn'd on the water—"

all the separate points of glitter in the scene which the poet
is describing become a single flame. Coleridge called this
unity in variety, and held that it is the basis of beauty. And,
of course, the opposite trick of putting two incongruous
things side by side is the stock in trade of humour. In all our
thinking, in science or outside it, we are always, all of us,
trying to find a unity in our human experiences.

The new theory which Dr. Einstein has published is called
a unified field theory. So it is meant to be another step in
drawing together the different modes of action of the physical
Universe into a single system. Ever since Newton took the
first step, back in the years of the Great Plague and the Fire of
London, this process has been going on, compacting and tying
together the concepts of science. We take these steps for
granted now, and few of us recall what a bold imaginative leap
each of them was in its own day. It is a commonplace to us
now that all matter is alike in being made of atoms ; that heat
is a form of energy, and is a movement of these atoms ; or that
an electric current carries energy in another way. More and
more we see the physical world linked and of a piece. We have
even become familiar in a lifetime with the two remarkable
unifying ideas which Einstein himself introduced when,
another marvellous boy, not much older than the young
Newton, he burst upon the world in 1905. One is that matter
and energy are forms of the same thing, and each can be
changed into the other. The second is that space and time are
not wholly separate things, but are connected parts of a single
reality.

But Einstein's new theory is concerned with still wider
topics. The two things which it sets out to link together are

gravitation and electro-magnetism. This ugly word " electro-magnetism " is itself the child of a long history of unification. Through much of the last century, physicists were finding the links between electricity and magnetism. The outstanding man in this work was Michael Faraday, who did most to show that each can be turned into the other. But Faraday did more than in this way to conceive the electric motor and the dynamo. He looked for links which no one else expected, such as the effect of magnetism on light ; and he found it. Nowadays when a scientist talks about electro-magnetism he is likely to be thinking above all of light and of radio waves. The beginning for this was made by Faraday. It was carried on in the wonderful theoretical work of a mathematician, James Clerk Maxwell, who showed that the same mathematical relations describe the behaviour of all these forms of energy.

Here then we have two great natural forms of energy. One is electro-magnetism, which includes electricity and magnetism, heat and light, infra-red and ultra-violet rays, radio waves, X-rays and even radio-active gamma rays, all in a single family obeying the same laws. The second is gravitation, not quite as Newton thought of it but in the less mechanical form which Einstein gave it in his theory of Relativity. These two are in effect the only large-scale physical forces which are still separate. Are they also connected ? People have often suspected it. Faraday himself gave a lecture, 100 years ago, with the title " On the Possible Relation of Gravity to Electricity ". In it he described his experiments to find a relation, which all failed ; nevertheless he said, " they do not shake my strong feeling of the existence of a relation ".

One likeness between light, say, and gravity is that they both spread out into space in the same way, evenly in every direction from the point where they began ; so that the strength of each falls off as the square of the distance from there. And then, of course, Einstein himself predicted in the theory of Relativity that light must be as it were attracted by matter, and was proved right spectacularly in recent eclipses.

Another likeness has only become certain in the last few years, because it is another consequence of the theory of Relativity. This is the discovery that gravitation does not get everywhere instantly, but travels in waves like electro-magnetic waves, and at the same speed, which is the speed of light. Clerk Maxwell was sure that light and electro-magnetism are allied when his mathematics showed that they travel at the same speed. So it is at least tempting to couple gravitation with them now that we find that that has the same speed too.

Let me put this to you in a simple picture. Suppose you had asked Shakespeare in the year 1600 this question : " What would happen on Earth if the Sun suddenly vanished into nothing ? " He would have answered, " It would get dark ". And if you pressed your question and asked " How soon ? " Shakespeare would say " At once, of course ". No one in 1600 knew that it takes time for light to reach us from the Sun. And Shakespeare would not have given a thought to the path of the Earth if the Sun vanished. Because few people then believed that the Earth goes round the Sun, and those who did had no notion what keeps it there.

Now suppose you had put the same question to Christopher Wren 100 years later, in 1700. By this time it was known that it takes time for light to travel. And Christopher Wren, who was a friend of Newton, of course also knew that it is the gravitational pull of the Sun which keeps the Earth on its path ; but he supposed that it reaches everywhere instantly. So he would answer, " If the Sun were suddenly to vanish into nothing now, the Earth would at once fly straight out of its orbit, but it wouldn't get dark for about eight minutes ".

And to-day ? We know that gravitation travels no faster than light and other waves. So to-day we should say, " If the Sun were suddenly to be snatched out of the sky, then the Earth would fly out of its orbit and it would get dark, both at the same moment . . . a little over eight minutes from now ! "

This is the kind of thought which has kept Einstein busy

for more than twenty years, looking for a unified theory which would at last forge the outstanding link between gravitation and electro-magnetism. He has not been afraid to publish one or two trial attempts ; but the theory which he has published now is plainly meant to be something more rounded and final. What does such a theory look like ? It is a set of mathematical relations which describe the field set up in space and time by the action of gravitation and electro-magnetism together. But then, what is the point of such a theory ? How does it affect our lives ? If Einstein is right, he will have brought together two strands in the structure of the world which had seemed separate. Every theory of this kind, which helps to unify our understanding of the structure of the world, is a major step forward, which re-shapes our thought and our future together. Clerk Maxwell in the eighteen-sixties took such a step when he linked light with electricity, and Darwin when, a little before, he gave unity to the vast tangle of living species. Their thoughts set off intellectual revolutions, and in time had such practical results as genetics and the radio set. The outstanding example for our generation is of course Einstein's own work. Nothing could have seemed more abstract than Einstein's formulation in 1905 of the Special Theory of Relativity ; yet in a hundred shapes, the concept of Relativity has dominated our thinking ever since. Nothing could have seemed more remote than Einstein's linking together in that paper of mass and energy. Yet on that innocent equation rests atomic energy of to-day, and with it the hopes and fears of our civilization.

So do not think of Einstein as an abstract thinker who pushes mathematical formulæ about. A little while ago, in a lecture room in Cambridge, I unexpectedly came across a blackboard covered with symbols, which had been carefully varnished and preserved. It was the board on which Einstein had lectured to us on this very subject nearly twenty years ago. But recalling the lecture suddenly, it was not the formulæ I re-membered. It was Einstein thinking aloud, vividly, physically :

I can hear him talking now—" What would happen in a lift ? " " What would happen in a rocket ? " It was typical of the man, simple, warm and wise, and each line an experiment and a challenge to thought. He who had done so much in many branches of science was looking then modestly for this one link which was still missing in large-scale physics, between the fields of force set up by gravitation and by electro-magnetism. It will be a long time before we shall know from actual physical experiment whether he has indeed now found it. Every scientist hopes that he has ; and that not only because it will round off one side of physics, and must have great practical results ; but because it will sum up the life-work of the man who has done more than any other to re-make and to knit together our understanding of the physical world.

Dr. Macpherson mentioned the possibility of life existing elsewhere in the universe, and perhaps you would like to ask an Astronomer Royal what he thinks about this possibility. *Does Life exist on other Planets ?* In his answer we shall find an outline of a recent popular theory on the origin of the solar system, together with a strong suggestion that this theory needs considerable modification. No theory on this subject seems to last very long. The latest is that the planets were born as the result of a terrific cosmic explosion in a starry companion of the sun.

SIR H. SPENCER JONES

LIFE ON OTHER WORLDS

IN the previous talks I have tried to take you for a survey of the planets in the solar system and of their satellites, in search for evidence of the existence of life. The results have not been encouraging to those of us who hoped to find clear evidence of life on other planets. For the most part we were able to say with certainty that life did not and could not exist. In Venus, however, we found a world where it seemed that life might be on the verge of coming into existence and where, conceivably, in the millions of years to come life may develop and flourish. Mars, on the other hand, appears to be a world in the sere and yellow leaf, where life may have flourished in ages long past but where conditions now seem to be such that, though plant life continues to exist, highly developed types of animal life are unlikely to be found.

This survey of mine has dealt only with the family of our Sun, which is merely an average star, one amongst the many stars, numbering some two hundred thousand millions or so, in our stellar universe. And that universe in turn is merely one amongst many millions of more or less similar island universes—each a gigantic system—scattered through space.

The question will inevitably be asked : What is the chance that life exists on some of the planets belonging to one or other of these innumerable stars in our own, or some other universe ? This is a difficult question to answer, because if such planetary systems exist, they are far beyond our range of vision. If the nearest known star, twenty-five million million miles away, had a planet belonging to it of the size of Jupiter—which is much the largest of the planets in our solar system—we should not be able to see it. The direct observation and study of planets belonging to other stars is therefore entirely out of the question.

But we can get some guidance from general considerations. First, what likelihood is there that other stars have families of planets at all ? We can give some sort of answer to this if we can discover how our Sun came to give birth to a family of planets. That is a problem to which a vast amount of thought and investigation has been given and it is one of the most difficult problems that astronomy has to attempt to solve. Unless I talked to you for hours I could not explain how one after another hypothesis that has been suggested has been shown to be incapable of explaining all the features of the solar system. In this game, knocking down the skittles proves to be much easier than setting them up.

Until quite recently there seemed to be one—and only one—hypothesis that could be made to work. According to this hypothesis, a few thousand million years ago another star passed close to the Sun. As it drew near, it raised, by gravitational attraction, a great tidal protuberance on the Sun—which, we must remember, is a gaseous body, not liquid or solid. This tidal wave we can think of as analogous to the heaping up of the ocean waters produced by the gravitational pull of the Moon, which we call the Tides, but on a much vaster scale. As the stranger star drew nearer, the tidal protuberance on the Sun became greater and greater until at length a long jet of matter was drawn out from the Sun. The stranger passed on its way, in its journey through space, not heeding the disturbance it had caused in the placid existence of the Sun. The tidal wave on the Sun subsided, but the evidence of the encounter has remained for ever in the matter drawn from the Sun. This broke up and condensed into the planets, of which our Earth is one.

Now, assuming for the moment that there is no flaw in the argument, we can estimate the probability of two stars approaching sufficiently close for planets to be born in this way. It is easy to see that the probability must be extremely small, because the stars are so far apart from one another. As I have already mentioned, the nearest star to us—so far as we know—is twenty-five million million miles away. We can get a better picture, perhaps, in this way : suppose we imagine six tennis balls to be flying about inside a hollow globe, the size of the Earth, 8000 miles in diameter. Then the chance of any two of these balls colliding with each other is equal to the chance of a near approach of two stars. It can be shown that it will only happen on the average about once in five thousand million years. If planets are born in the way we have supposed, it must follow that planetary systems are not the rule but very much the exception, and that in our vast stellar universe there can be but few stars, in addition to our Sun, which have systems of planets attached to them.

Recently a flaw has been found in the argument, however. In order to make the theory work it appears to be necessary to suppose not merely that another star passed near the Sun, but also that the Sun was at that time a member of a twin system. As a matter of fact twin stars revolving round each other under their mutual gravitation are quite common : probably one star in every five is a twin. Still, this additional assumption makes the theory appear somewhat more artificial. However, the solar system has come into existence somehow, and if we can find one way in which its existence can be accounted for, we must consider it seriously, in the absence of any alternative explanation, as the probably correct explanation. But the upshot of this modification of the theory is that we must conclude that planetary systems are even rarer than we thought they were a few years ago. Nevertheless, when we consider the vast number of separate universes, even though we assume that there can be at most only several stars with systems of planets in each of these universes, yet it appears that the total number of planetary systems is considerable.

But the existence of other planetary systems, though a necessary condition for life to exist elsewhere in the Universe, is not a sufficient condition. In any planetary system everything seems to be weighted against the possibility of the existence of life. If the planet is very near its parent sun, it will be too hot for life to be possible ; if it is very far away, it will be too cold. If it is much smaller than the Earth, it will have been unable to retain any atmosphere. If it is much larger, it will have retained too much atmosphere, for when the gravitational attraction is so great that hydrogen cannot escape from the atmosphere the formation of the poisonous gases, ammonia and marsh gas—which we found in the atmosphere of Jupiter and Saturn—appears to be almost inevitable.

It seems, in fact, that for life to exist on a planet anywhere in the Universe, the planet must not differ greatly in size and weight from our Earth. This condition being satisfied, the planet must be placed at such a distance from its parent sun that it is neither very hot nor very cold. If the parent sun is much hotter than our Sun, the planet would have to be further away from its parent than we are from the Sun; if it is much cooler than the Sun, the planet would have to be much nearer. A somewhat exact adjustment of two factors, the size of the planet and its distance from its parent sun, thus seems to be essential if life is to be possible on the planet. It is not enough to have either factor satisfied without the other.

To sum up this argument, the conditions needed for birth to be given to a planetary system are so exceptional that amongst the vast number of stars in any one stellar universe we should expect to find only a very limited number that have a family of planets; and amongst these families of planets there cannot be more than a small proportion where the conditions are suitable for life to exist. Life elsewhere in the Universe is therefore the exception and not the rule. If we could travel through the Universe and survey each star in turn, we should not find life here, there, and everywhere. To find a needle in a haystack would be easy in comparison with the task of finding another world where there is life. But is not this typical of the universal prodigality of Nature?

Does it follow that our Earth is unique, the only home of life in the Cosmos as a whole—the Universe of Universes? I cannot think that this can be so. In the region of space that can be surveyed with the most powerful telescope yet built there are some sixty million separate universes; even if, in each of these universes, there are not more than two or three dozen stars with families of planets, the total number of planetary systems within the relatively small region of space that we can survey must all the same be very great, a few thousand millions. If we suppose the proportion of planets on which life can exist is not more than one in a thousand, or even one in a million—and let me remark, the consideration of our solar system suggests that this is a considerable underestimate—the total number of worlds throughout the Universe where conditions are suitable for life to exist, must be rather large.

If we hold, as I think we must, that, where conditions are suitable for life, life will somehow develop, it follows that life must exist on many other worlds. It may not be life such as we are familiar with—probably as a result of some slight difference in conditions, entirely different forms of life will develop, though there may well be intelligent life. Our little Earth is not unique ; in my opinion the whole purpose of creation has not been centred on this one small world.

And now, as a relief from the contemplation of long ages and immense distances, vast numbers and colossal explosions, let us come back to the earth and its next-door neighbour, and ask *How does the Moon affect human life ?* We are studying science, so we must leave aside its effect on young lovers and poets. Its more practical consequences are outlined by Robert Ball, who used to be one of the most popular of writers on Astronomy.

Although Sir Robert, in his lecture, ascribes the origin of the striking craters on the moon's surface (some of them 150 miles across) to volcanic action, it must be said that some astronomers prefer to explain them as due to the impact of meteors on the moon while it was still in a plastic condition, while others attribute them to the bursting of bubbles on the molten surface before it solidified. This question is still open.

SIR ROBERT BALL

THE MOON

IF the moon were suddenly struck out of existence, we should be immediately apprised of the fact by a wail from every seaport in the kingdom. From London and from Liverpool we should hear the same story—the rise and fall of the tide had almost ceased. The ships in dock could not get out ; the ships outside could not get in ; and the maritime commerce of the world would be thrown into dire confusion.

The moon is the principal agent in causing the daily ebb and flow of the tide, and this is the most important work which our satellite has to do. The fleets of fishing boats around the coasts time their daily movements by the tide, and are largely indebted to the moon for bringing them in and out of harbour. Experienced sailors assure us that the tides are of the utmost service to navigation.

Who is there that has not watched, with admiration, the beautiful series of changes through which the moon passes every month ? We first see her as an exquisite crescent of pale light in the western sky after sunset. If the night is fine, the rest of the moon is visible inside the crescent, being faintly illumined by light reflected from our own earth. Night after night she moves further and further to the east, until she becomes full, and rises about the same time that the sun sets. From the time of the full the disc of light begins to diminish until the last quarter is reached. Then it is that the moon is seen high in the heavens in the morning. As the days pass by, the crescent shape is again assumed. The crescent wanes thinner and thinner as the satellite draws closer to the sun. Finally she becomes lost in the overpowering light of the sun, again to emerge as the new moon, and again to go through the same cycle of changes.

The brilliance of the moon arises solely from the light of the sun, which falls on the not self-luminous substance of the moon. Out of the vast flood of light which the sun pours forth with such prodigality into space the dark body of the moon intercepts a little, and of that little it reflects a small fraction to illuminate the earth. The moon sheds so much light, and seems so bright, that it is often difficult at night to remember that the moon has no light except what falls on it from the sun. Nevertheless, the actual surface of the brightest full moon is perhaps not much brighter than the streets of

London on a clear sunshiny day. A very simple observation will suffice to show that the moon's light is only sunlight. Look some morning at the moon in daylight, and compare the moon with the clouds. The brightness of the moon and of the clouds are directly comparable, and then it can be readily comprehended how the sun which illuminates the clouds has also illumined the moon. An attempt has been made to form a comparative estimate of the brightness of the sun and the full moon. If 600,000 full moons were shining at once, their collective brilliancy would equal that of the sun.

The beautiful crescent moon has furnished a theme for many a poet. Indeed, if we may venture to say so, it would seem that some poets have forgotten that the moon is not to be seen every night. A poetical description of evening is almost certain to be associated with the appearance of the moon in some phase or other. We may cite one notable instance in which a poet, describing an historical event, has enshrined in exquisite verse a statement which cannot be correct. Every child who speaks our language has been taught that the burial of Sir John Moore took place—

" By the struggling moonbeams' misty light."

There is an appearance of detail in this statement which wears the garb of truth. We are not inclined to doubt that the night was misty, nor as to whether the moonbeams had to struggle into visibility ; the question at issue is a much more funda-mental one. We do not know who was the first to raise the point as to whether any moon shone on that memorable event at all or not ; but the question having been raised, the Nautical Almanac immediately supplies an answer. From it we learn in language whose truthfulness constitutes its only claim to be poetry that the moon was new at one o'clock in the morning of the day of the battle of Corunna (January 16, 1809). The ballad evidently implies that the funeral took place on the night following the battle. We are therefore assured that the moon can hardly have been a day old when the hero was consigned to his grave. But the moon in such a case is practically invisible, and yields no appreciable moonbeams at all, misty or otherwise. Indeed, if the funeral took place at the " dead of night ", as the poet asserts, then the moon must have been far below the horizon at the time.

In alluding to this and similar instances, Mr. Nasmyth gives a word of advice to authors or to artists who desire to bring the moon on a scene without knowing as a matter of fact that our satellite was actually present. He recommends them to follow the example of Bottom in *A Midsummer-Night's Dream* and consult " a calendar, a calendar ! Look in the almanac ; find out moonshine, find out moonshine ! "

Among the countless host of celestial bodies—the sun, the moon, the planets, and the stars—our satellite enjoys one special claim on our attention. The moon is our nearest permanent neighbour. It is just possible that a comet may occasionally approach the earth more closely than the moon, but with this exception the other celestial bodies are generally hundreds of thousands, or even many millions, of times further from us than the moon.

It is also to be observed that the moon is one of the smallest visible objects which the heavens contain. Every one of the thousands of stars that can be seen with the unaided eye is enormously larger than our satellite. The brilliance and apparent vast proportions of the moon arise from the fact that it is only 239,000 miles away, which is a distance almost immeasurably small when compared with the distances between the earth and the stars.

During a long voyage, and perhaps in critical circumstances, the moon will often render invaluable information to the sailor. To navigate a ship, suppose from Liverpool to China, the captain must frequently determine the precise position which his ship then occupies. If he could not do this, he would never find his way across the trackless ocean. Observations of the sun give him his latitude and tell him his local time, but the captain further requires to know the Greenwich time before he can place his finger at a point of the chart and say, " My ship is here ". To ascertain the Greenwich time the ship carries a chronometer which has been carefully rated before starting, and, as a precaution, two or three chronometers are usually provided to guard against the risk of error. An unknown error of a minute in the chronometer might perhaps lead the vessel fifteen miles from its proper course.

It is important to have the means of testing the chronometers during the progress of the voyage ; and it would be a great convenience if every captain, when he wished, could actually consult some infallible standard of Greenwich time. We want, in fact, a Greenwich clock which may be visible over the whole globe. There is such a clock ; and, like any other clock, it has a face on which certain marks are made, and a hand which travels round that face. The great clock at Westminster shrinks into insignificance when compared with the mighty clock which the captain uses for setting his chronometer. The face of this stupendous dial is the face of the heavens. The numbers engraved on the face of a clock are replaced by the twinkling stars ; while the hand which moves over the dial is the beautiful moon herself. When the captain desires to test his chronometer, he measures the distance of the moon from a neighbouring star. In the Nautical Almanac he finds the Greenwich time at which the moon was three degrees from the star. Comparing this with the indications of the chronometer, he finds the required correction. In recent years ships carry more chronometers, so the lunar method has fallen into disuse.

There is one widely credited myth about the moon which must be regarded as devoid of foundation. The idea that our satellite and the weather bear some relation has no doubt been entertained by high authority, and appears to be an article in the belief of many an excellent mariner. Careful comparison between the state of the weather and the phases of the moon has, however, quite discredited the notion that any connexion of the kind does really exist.

The lunar landscapes are excessively weird and rugged. They always remind us of sterile deserts, and we cannot fail to notice the absence of grassy plains or green forests such as we are familiar with on our globe. In some respects the moon is not very differently circumstanced from the earth. Like it, the moon has the pleasing alternations of day and night, though the day in the moon is as long as twenty-nine of our days, and the night of the moon is as long as twenty-nine of our nights. We are warmed by the rays of the sun ; so, too, is the moon ; but, whatever may be the temperature during the long day on the moon, it seems certain that the cold of the lunar night would transcend that known in the bleakest regions of our earth. The amount of heat radiated to us by the moon has been investigated by Lord Rosse, and more recently by Professor Langley. Though every point on the moon's surface is exposed to the sunlight for a fortnight, without any interruption, the actual temperature to which the soil is raised cannot be a high one. The moon does not, like the earth, possess a warm blanket, in the shape of an atmosphere, which can keep in and accumulate the heat received.

Even our largest telescopes can tell nothing directly as to whether life can exist on the moon. The mammoth trees of California might be growing on the lunar mountains, and elephants might be walking about on the plains, but our telescopes could not show them. The smallest object that we can see on the moon must be about as large as a good-sized cathedral, so that organized beings resembling in size any that we are familiar with, if they existed, could not make themselves visible as telescopic objects.

We are therefore compelled to resort to indirect evidence as to whether life would be possible on the moon. We may say at once that astronomers believe that life, as we know it, could not exist. Among the necessary conditions of life, water is one of the first. Take every form of vegetable life, from the lichen which grows on the rock to the giant tree of the forest, and we find the substance of every plant contains water, and could not exist without it. Nor is water less necessary to the existence of animal life. Deprived of this element, all organic life, the life of man himself, would be inconceivable.

Unless, therefore, water be present in the moon, we shall be bound to conclude that life, as we know it, is impossible. If anyone stationed on the moon were to look at the earth through a telescope, would he be able to see any water here? Most undoubtedly he would. He would see the clouds and he would notice their incessant changes, and the clouds alone would be almost conclusive evidence of the existence of water. An astronomer on the moon would also see our oceans as coloured surfaces, remarkably contrasted with the land, and he would perhaps frequently see an image of the sun, like a brilliant star, reflected from some smooth portion of the sea. In fact, considering that much more than half of our globe is covered with oceans, and that most of the remainder is liable to be obscured by clouds, the lunar astronomer in looking at our earth would often see hardly anything but water in one form or another. Very likely he would come to the conclusion that our globe was fitted to be a residence for only amphibious animals.

But when we look at the moon with our telescopes we see no direct evidence of water. Close inspection shows that the so-called lunar seas are deserts, often marked with small craters and rocks. The telescope reveals no seas and no oceans, no lakes and no rivers. Nor is the grandeur of the moon's scenery ever impaired by clouds over her surface. Whenever the moon is above our horizon, and terrestrial clouds are out of the way, we can see the features of our satellite's surface with distinctness. There are no clouds in the moon; there are not even the mists or the vapours which invariably arise wherever water is present, and therefore astronomers have been led to the conclusion that the surface of the globe which attends the earth is a sterile and a waterless desert.

Another essential element of organic life is also absent from the moon. Our globe is surrounded with a deep clothing of air resting on the surface, and extending above our heads to the height of about 200 or 300 miles. We need hardly say how necessary air is to life, and therefore we turn with interest to the question as to whether the moon can be surrounded with an atmosphere. Let us clearly understand the problem we are about to consider. Imagine that a traveller started from the earth on a journey to the moon ; as he proceeded, the air would gradually become more and more rarefied, until at length, when he was a few hundred miles above the earth's surface, he would have left the last perceptible traces of the earth's envelope behind him. By the time he had passed completely through the atmosphere he would have advanced only a very small fraction of the whole journey of 239,000 miles, and there would still remain a vast void to be traversed before the moon would be reached. If the moon were enveloped in the same way as the earth, then, as the traveller approached the end of his journey, and came within a few hundred miles of the moon's surface, he would meet again with traces of an atmosphere, which would gradually increase in density until he arrived at the moon's surface. The traveller would thus have passed through one stratum of air at the beginning of his journey, and through another at the end, while the main portion of the voyage would have been through space more void than that to be found in the exhausted receiver of an air-pump.

Such would be the case if the moon were coated with an atmosphere like that surrounding our earth. But what are the facts ? The traveller as he drew near the moon would seek in vain for air to breathe at all resembling ours. It is possible that close to the surface there are faint traces of some gaseous material surrounding the moon, but it can only be equal to a very small fractional part of the ample clothing which the earth now enjoys. For all purposes of respiration, as we understand the term, we may say that there is no air on the moon, and an inhabitant of our earth transferred thereto would be as certainly suffocated as he would be in the middle of space.

The absence of air and of water from the moon explains the sublime ruggedness of the lunar scenery. We know that on the earth the action of wind and of rain, of frost and of snow, is constantly tending to wear down our mountains and reduce their asperities. No such agents are at work on the moon. Volcanoes sculptured the surface into its present condition, and, though they have ceased to operate for ages, the traces of their handiwork seem nearly as fresh to-day as they were when the mighty fires were extinguished.

"The cloud-capped towers, the gorgeous palaces, the solemn temples" have but a brief career on earth. It is chiefly the incessant action of water and of air that makes them vanish like the "baseless fabric of a vision". On the moon these causes of disintegration and of decay are all absent, though perhaps the changes of temperature in the transition from lunar day to lunar night would be attended with expansions and contractions that might compensate in some slight degree for the absence of more potent agents of dissolution.

It seems probable that a building on the moon would remain for century after century just as it was left by the builders. There need be no glass in the windows, for there is no wind and no rain to keep out. There need not be fireplaces in the rooms, for fuel cannot burn without air. Dwellers in a lunar city would find that no dust could arise, no odours be perceived, no sounds be heard.

Man is a creature adapted for life under circumstances which are very narrowly limited. A few degrees of temperature more or less, a slight variation in the composition of air, the precise suitability of food, make all the difference between health and sickness, between life and death. Looking beyond the moon, into the length and breadth of the universe, we find countless celestial globes with every conceivable variety of temperature and of constitution. Amid this vast number of worlds with which space is tenanted, are there any inhabited by living beings? To this great question science can make no response: we cannot tell. Yet it is impossible to resist a conjecture. We find our earth teeming with life in every part. We find life under the most varied conditions that can be conceived. It is met with under the burning heat of the tropics and in the everlasting frosts at the poles. We find life in caves where not a ray of light ever penetrates. Nor is it wanting in the depths of the ocean, at the pressure of tons on the square inch. Whatever may be the external circumstances, Nature generally provides some form of life to which those circumstances are congenial.

It is not at all probable that among the million spheres of the universe there is a single one exactly like our earth—like it in the possession of air and of water, like it in size and in composition. It does not seem probable that a man could live for one hour on any body in the universe except the earth, or that an oak-tree could live in any other sphere for a single season. Men can dwell on the earth, and oak-trees can thrive therein, because the constitutions of the man and of the oak are specially adapted to the particular circumstances of the earth.

Could we obtain a closer view of some of the celestial bodies, we should probably find that they, too, teem with life, but with life specially adapted to the environment—life in forms strange and weird ; life far stranger to us than Columbus found it to be in the New World when he first landed there. Life, it may be, stranger than ever Dante described or Doré sketched. Intelligence may also have a home among those spheres no less than on the earth. There are globes greater and globes less—atmospheres greater and atmospheres less. The truest philosophy on this subject is crystallized in the language of Tennyson :—

> " This truth within thy mind rehearse,
> That in a boundless universe
> Is boundless better, boundless worse.
>
> Think you this mould of hopes and fears
> Could find no statelier than his peers
> In yonder hundred million spheres ? "

We conclude this section on the Universe by taking a brief glance at the way in which the surface of our Earth has been sculptured. *How can the puzzling story of the rocks be deciphered?*

Sir Charles Lyell took as his guide the careful observation of what was happening day by day on mountains and in river valleys, in deltas and lakes and on the sea shore, and with this help he explained how the different kinds of rock had been formed.

SIR CHARLES LYELL

THE CRUST OF THE EARTH

OF what materials is the earth composed, and in what manner are these materials arranged? These are the first enquiries with which Geology is occupied, a science which derives its name from the Greek, *ge*, the earth, and *logos*, a discourse. Previously to experience we might have imagined that investigations of this kind would relate exclusively to the mineral kingdom, and to the various rocks, soils, and metals, which occur upon the surface of the earth, or at various depths beneath it. But, in pursuing such researches, we soon find ourselves led on to consider the successive changes which have taken place in the former state of the earth's surface and interior, and the causes which have given rise to these changes ; and, what is still more singular and unexpected, we soon become engaged in researches into the history of the animate creation, or of the various tribes of animals and plants which have, at different periods of the past, inhabited the globe.

All are aware that the solid parts of the earth consist of distinct substances, such as clay, chalk, sand, limestone, coal, slate, granite, and the like ; but previously to observation it is commonly imagined that all these have remained from the first in the state in which we now see them—that they were created in their present form, and in their present position. The geologist soon comes to a different conclusion, discovering proofs that the external parts of the earth were not all produced in the beginning of things in the state in which we now behold them, nor in an instant of time. On the contrary, he can show that they have acquired their actual configuration and condition gradually, under a great variety of circumstances, and at successive periods, during each of which distinct races of living beings have flourished on the land and in the waters, the remains of these creatures still lying buried in the crust of the earth.

By the " earth's crust " is meant that small portion of the exterior of our planet which is accessible to human observation. It comprises not merely all of which the structure is laid open in mountain precipices, or in cliffs overhanging a river or the sea, or whatever the miner may reveal in artificial excavations ; but the whole of that outer covering of the planet on which we are enabled to reason by observations made at or near the surface. These reasonings may extend to a depth of several miles, perhaps ten miles ; and even then it may be said, that such a thickness is no more than one four-hundredth part of the distance from the surface to the centre. The remark is just ; but although the dimensions of such a crust are, in truth, insignificant when compared to the entire globe, yet they are vast, and of magnificent extent in relation to man, and to the organic beings which people our globe. Referring to this standard of magnitude, the geologist may admire the ample limits of his domain, and admit, at the same time, that not only the exterior of the planet, but the entire earth, is but an atom in the midst of the countless worlds surveyed by the astronomer.

The materials of this crust are not thrown together confusedly, but distinct mineral masses, called rocks, are found to occupy definite spaces, and to exhibit a certain order of arrangement. The term *rock* is applied indifferently by geologists to all these substances, whether they be soft or stony, for clay and sand are included in the term, and some have even brought peat under this denomination.

The most natural and convenient mode of classifying the

various rocks which compose the earth's crust is to refer, in the first place to their origin, and in the second to their relative age. I shall therefore begin by endeavouring briefly to explain how all rocks may be divided into four great classes by reference to their different origin, or, in other words, by reference to the different circumstances and causes by which they have been produced.

The first two divisions, which will at once be understood as natural, are the aqueous and volcanic, or the products of watery and those of igneous action at or near the surface.

AQUEOUS ROCKS

The aqueous rocks, sometimes called the sedimentary, or fossiliferous, cover a larger part of the earth's surface than any others. They consist chiefly of mechanical deposits (pebbles, sand, and mud) but are partly of chemical and some of them of organic origin, especially the limestones. These rocks are *stratified*, or divided into distinct layers, or strata. The term *stratum* means simply a bed, or anything spread out or *strewed* over a given surface ; and we infer that these strata have been generally spread out by the action of water, from what we daily see taking place near the mouths of rivers, or on the land during temporary inundations. For, whenever a running stream charged with mud or sand has its velocity checked, as when it enters a lake or sea, or overflows a plain, the sediment, previously held in suspension by the motion of the water, sinks by its own gravity, to the bottom. In this manner, layers of mud and sand are thrown down one upon another.

If we drain a lake which has been fed by a small stream, we frequently find at the bottom a series of deposits, disposed with considerable regularity, one above the other ; the uppermost, perhaps, may be a stratum of peat, next below a more dense and solid variety of the same material : still lower a bed of shell-marl, alternating with peat or sand, and then other beds of marl, divided by layers of clay. Now, if a second pit be sunk through the same continuous lacustrine *formation* at some distance from the first, nearly the same series of beds is commonly met with, yet with slight variations ; some, for example, of the layers of sand, clay, or marl, may be wanting, one or more of them having thinned out and given place to others, or sometimes one of the masses first examined is observed to increase in thickness to the exclusion of other beds.

The term " formation ", which I have used in the above explanation, expresses in geology any assemblage of rocks which have some character in common, whether of origin, age, or composition. Thus we speak of stratified and unstratified, freshwater and marine, aqueous and volcanic, ancient and modern, metalliferous and non-metalliferous formations.

In the estuaries of large rivers, such as the Ganges and the Mississippi, we may observe, at low water, phenomena analogous to those of the drained lakes above mentioned, but on a grander scale, and extending over areas several hundred miles in length and breadth. When the periodical inundations subside, the river hollows out a channel to the depth of many yards through horizontal beds of clay and sand, the ends of which are seen exposed in perpendicular cliffs. These beds vary in their mineral composition, or colour, or in the fineness or coarseness of their particles, and some of them are occasionally characterized by containing driftwood. At the junction of the river and the sea, especially in lagoons nearly separated by sand bars from the ocean, deposits are often formed in which brackish and salt-water shells are included. . . .

When beds of sand, clay, and marl, containing shells and vegetable matter, are found arranged in a similar manner in the interior of the earth, we ascribe to them a similar origin ; and the more we examine their characters in minute detail, the more exact do we find the resemblance. Thus, for example, at various heights and depths in the earth, and often far from seas, lakes, and rivers, we meet with layers of rounded pebbles composed of flint, limestone, granite, or other rocks, resembling the shingles of a sea-beach or the gravel in a torrent's bed.

Such layers of pebbles frequently alternate with others formed of sand or fine sediment, just as we may see in the channel of a river descending from hills bordering a coast, where the current sweeps down at one season coarse sand and gravel, while at another, when the waters are low and less rapid, fine mud and sand alone are carried seaward.

If a stratified arrangement, and the rounded form of pebbles are alone sufficient to lead us to the conclusion that certain rocks originated under water, this opinion is further confirmed by the distinct and independent evidence of *fossils*, so abundantly included in the earth's crust. By a *fossil* is meant any body, or the traces of the existence of any body, whether animal or vegetable, which has been buried in the earth by natural causes. Now the remains of animals, especially of aquatic species, are found almost everywhere embedded, in stratified rocks, and sometimes, in the case of limestone, they are in such abundance as to constitute the entire mass of the rock itself. Shells and corals are the most frequent, and with them are often associated the bones and teeth of fishes, fragments of wood, impressions of leaves, and other organic substances. Fossil shells of forms such as now abound in the sea are met with far inland, both near the surface and at great depths below it. They occur at all heights above the level of the ocean, having been observed at elevations of more than 8000 feet in the Pyrenees, 10,000 in the Alps, 13,000 in the Andes, and above 18,000 feet in the Himalaya.

These shells belong mostly to marine testacea, but in some places exclusively to forms characteristic of lakes and rivers. Hence it is concluded that some ancient strata were deposited at the bottom of the sea, and others in lakes and estuaries.

We have now pointed out one great class of rocks, which, however they may vary in mineral composition, colour, grain, or other characters, external and internal, may nevertheless be grouped together as having a common origin. They have all been formed under water, in the same manner as modern accumulations of sand, mud, shingle, banks of shells, reefs of coral, and the like, and are all characterized by stratification or fossils or by both.

VOLCANIC ROCKS

The division of rocks which we may next consider are the volcanic, or those which have been produced at or near the surface whether in ancient or modern times, not by water but by the action of fire or subterranean heat. These rocks are for the most part unstratified, and are devoid of fossils. They are more partially distributed than aqueous formations, at least in respect to horizontal extension. Among those parts of Europe where they exhibit characters not to be mistaken, I may mention not only Sicily and the country round Naples, but Auvergne, Velay, and Vivarais, now the departments of Puy-de-Dôme, Haute-Loire, and Ardêche, towards the centre and south of France, in which are several hundred conical hills having the forms of modern volcanoes, with craters more or less perfect on many of their summits. These cones are composed, moreover, of lava, sand, and ashes, similar to those of active volcanoes. Streams of lava may sometimes be traced from the cones into the adjoining valleys, where they have choked up the ancient channels of rivers with solid rock, in the same manner as some modern flows of lava in Iceland have been known to do, the rivers either flowing beneath or cutting out a narrow passage on one side of the lava. Although none of these French volcanoes have been in activity within the period of history or tradition, their forms are often very perfect. Some, however, have been compared to the mere skeletons of volcanoes, the rains and torrents having washed their sides, and removed all the loose sand and scoriæ, leaving only the harder and more solid materials. By this erosion, and by earthquakes, their internal structure has occasionally been laid open to view, in fissures and ravines ; and we then behold not only many successive beds and masses of porous lava, sand, and scoriæ, but also perpendicular walls, or *dikes*, as they are called, of volcanic rock, which have burst through the other materials. Such dikes are also observed in the structure of Vesuvius, Etna, and other active volcanoes. They have been formed by the pouring of melted matter, whether from above or below, into open fissures, and they commonly traverse deposits of *volcanic tuff*, a substance produced by the showering down from the air, or incumbent waters, of sand and cinders, first shot up from the interior of the earth by the explosions of volcanic gases.

Besides the parts of France above alluded to, there are other countries, as the north of Spain, the south of Sicily, the Tuscan territory of Italy, the lower Rhenish provinces, and Hungary, where spent volcanoes may be seen, still preserving in many cases a conical form, and having craters and often lava-streams connected with them.

There are also other rocks in England, Scotland, Ireland, and almost every country in Europe, which we infer to be of igneous origin, although they do not form hills with cones and craters. Thus, for example, we feel assured that the rock of Staffa, and that of the Giant's Causeway, called basalt, is volcanic, because it agrees in its columnar structure and mineral composition with streams of lava which we know to have flowed from the craters of volcanoes. We find also similar basaltic and other igneous rocks associated with beds of *tuff* in various parts of the British Isles and forming *dikes*, such as have been spoken of; and some of the strata through which these dikes cut are occasionally altered at the point of contact, as if they had been exposed to the intense heat of melted matter.

The absence of cones and craters, and long narrow streams of superficial lava, in England and many other countries, is principally to be attributed to the eruptions having been submarine, just as a considerable proportion of volcanoes in our own times burst out beneath the sea. But this question must be enlarged upon more fully in the chapters on igneous rocks, in which it will also be shown that as different sedimentary formations, containing each their characteristic fossils, have been deposited at successive periods, so also volcanic sand and scoriæ have been thrown out, and lavas have flowed over the land or bed of the sea, at many different epochs, or have been injected into fissures ; so that the igneous as well as the aqueous rocks may be classed as a chronological series of monuments, throwing light on a succession of events in the history of the earth.

PLUTONIC ROCKS

We have now pointed out the existence of two distinct orders of mineral masses, the aqueous and the volcanic : but if we examine a large portion of a continent, especially if it contain within it a lofty mountain range, we rarely fail to discover two other classes of rocks, very distinct from either of those above alluded to, and which we can neither assimilate to deposits such as are now accumulated in lakes or seas, nor to those generated by ordinary volcanic action. The members of both these divisions of rocks agree in being highly crystalline and destitute of organic remains. The rocks of one division have been called plutonic, comprehending all the granites and certain porphyries, which are nearly allied in some of their characters to volcanic formations. The members of the other class are stratified and often slaty, and have been called by some the *crystalline schists,* in which group are included gneiss, micaceous-schist (or mica-slate), hornblende-schist, statuary marble, the finer kinds of roofing slate, and other rocks afterwards to be described.

As it is admitted that nothing strictly analogous to these crystalline productions can now be seen in the progress of formation on the earth's surface, it will naturally be asked on what data we can find a place for them in a system of classification founded on the origin of rocks. I cannot, in reply to this question, pretend to give the student, in a few words, an intelligible account of the long chain of facts and reasonings from which geologists have been led to infer the nature of the

rocks in question. The result, however, may be briefly stated.
All the various kinds of granites which constitute the plutonic
family are supposed to be of igneous or aqueo-igneous origin,
and to have been formed under great pressure, at a considerable depth in the earth, or sometimes, perhaps, under a certain
weight of incumbent ocean. Like the lava of volcanoes, they
have been melted, and afterwards cooled and crystallized, but
with extreme slowness, and under conditions very different
from those of bodies cooling in the open air. Hence they differ
from the volcanic rocks, not only by their more crystalline
texture, but also by the absence of tuffs and breccias, which are
the products of eruptions at the earth's surface, or beneath
seas of inconsiderable depth. They differ also by the absence
of pores or cellular cavities, to which the expansion of the
entangled gases gives rise in ordinary lava.

METAMORPHIC OR STRATIFIED CRYSTALLINE ROCKS

The fourth and last great division of rocks are the crystalline
strata and slates, or schists, called gneiss, mica-schist, clay-
slate, chlorite schist, marble, and the like, the origin of which
is more doubtful than that of the other three classes. They
contain no pebbles, or sand, or scoriæ, or angular pieces of
embedded stone, and no traces of organic bodies, and they are
often as crystalline as granite, yet are divided into beds, corresponding in form and arrangement to those of sedimentary
formations, and are therefore said to be stratified. The beds
sometimes consist of an alternation of substances varying in
colour, composition, and thickness, precisely as we see in
stratified fossiliferous deposits. According to the Huttonian
theory, which I adopt as the most probable, and which will be
afterwards the more fully explained, the materials of these
strata were originally deposited from water in the usual form
of sediment, but they were subsequently so altered by subterranean heat as to assume a new texture. It is demonstrable,
in some cases at least, that such a complete conversion has

actually taken place, fossiliferous strata having exchanged an earthy for a highly crystalline texture for a distance of a quarter of a mile from their contact with granite. In some cases, dark limestones, replete with shells and corals, have been turned into white statuary marble, and hard clays, containing vegetable or other remains, into slates called mica-schist or hornblende-schist, every vestige of the organic bodies having been obliterated.

Although we are in a great degree ignorant of the precise nature of the influence exerted in these cases, yet it evidently bears some analogy to that which volcanic heat and gases are known to produce; and the action may be conveniently called plutonic, because it appears to have been developed in those regions where plutonic rocks are generated, and under similar circumstances of pressure and depth in the earth. Intensely heated water or steam permeating stratified masses under great pressure have no doubt played their part in producing the crystalline texture and other changes, and it is clear that the transforming influence has often pervaded entire mountain masses of strata.

In accordance with the hypothesis above alluded to, I proposed the term " Metamorphic " for the altered strata, a term derived from meta, *trans*, and morphe, *forma*.

Hence there are four great classes of rocks considered in reference to their origin,—the aqueous, the volcanic, the plutonic, and the metamorphic. In the course of this work it will be shown that portions of each of these four distinct classes have originated at many successive periods. They have all been produced contemporaneously, and may even now be in the progress of formation on a large scale. It is not true, as was formerly supposed, that all granites, together with the crystalline or metamorphic strata, were first formed, and therefore entitled to be called " primitive ", and that the aqueous and volcanic rocks were afterwards superimposed, and should therefore rank as secondary in the order of time. This idea was adopted in the infancy of the science, when all formations, whether stratified or unstratified, earthy or crystalline, with or without fossils, were alike regarded as of aqueous origin. At that period it was naturally argued that the foundation must be older than the superstructure ; but it was afterwards discovered that this opinion was by no means in every instance a legitimate deduction from facts ; for the inferior parts of the earth's crust have often been modified, and even entirely changed, by the influence of volcanic and other subterranean causes, while superimposed formations have not been in the slightest degree altered. In other words, the destroying and renovating processes have given birth to new rocks below, while those above, whether crystalline or fossiliferous, have remained in their ancient condition. . . .

From what has now been said, the reader will understand that each of the four great classes of rocks may be studied under two distinct points of view ; first, they may be studied simply as mineral masses deriving their origin from particular causes, and having a certain composition, form, and position in the earth's crust, or other characters both positive and negative, such as the presence or absence of organic remains. In the second place, the rocks of each class may be viewed as a grand chronological series of monuments, attesting a succession of events in the former history of the globe and its living inhabitants.

BOOK III
MATTER AND ENERGY

MATTER AND ENERGY

THE Greek idea that all substances consist of a few simple elements united in different proportions persisted throughout the Middle Ages, and led to a widespread belief in the " sacred art " of alchemy. The alchemists chased two will-o'-the-wisps : an *elixir vitæ* which would cure all illnesses, and a means of transmuting base metals into gold.

From their experiments came a host of vague and fantastic writings and recipes, but also from their work the science of chemistry was born. For many centuries it struggled vainly to free itself from its swaddling clothes of astrology and wrong ideas. Even in the early eighteenth century Stahl added a further encumbrance with his theory of " phlogiston "—a substance with negative weight which escaped in the flames when a body burnt. But delivery was at hand. Black, Cavendish, and Priestley demolished the phlogiston theory, and Lavoisier placed the infant science in an atmosphere where it could grow by his series of careful experiments on different kinds of " air ". *What led Lavoisier to think that combustion and even the " flame of life " could be explained along ordinary chemical lines ?* He gave the answer in his Easter lecture to the Académie des Sciences in 1775.

A. L. LAVOISIER

THE DIFFERENT KINDS OF AIR

ARE there different kinds of air ? . . . Are the different airs that nature offers us, or that we succeed in making, exceptional substances, or are they modifications of atmospheric air ? Such are the principal subjects embraced in my scheme of work, the development of which I propose to submit to the Academy. But since the time devoted to our public meetings does not allow me to treat any one of these questions in full, I will confine myself to-day to one particular case, and will only show that the principle which unites with metals during calcination is nothing else than the healthiest and purest part of air, so that if air, after entering into combination with a metal, is set free again, it emerges in an eminently respirable condition, more suited than atmospheric air to support ignition and combustion.

The majority of metallic calces are only reduced, that is to say, only return to the metallic condition, by immediate contact with a carbonacecus material, or with some substance containing what is called *phlogiston*. The charcoal that one uses is entirely destroyed during the operation when the amount is in suitable proportion, whence it follows that the air set free from metallic reductions with charcoal is not simple ; it is in some way the result of the combination of the elastic fluid set free from the metal and that set free from the charcoal ; thus, though this fluid is obtained in the state of fixed air, it is not justifiable to conclude that it existed in this state in the metallic calx before its combination with the carbon.

These reflections made me feel how essential it was—in order to unravel the mystery of the reduction of metallic calces—to perform all my experiments on calces which can be reduced without addition of charcoal. . . .

Precipitated mercury, which is nothing else than a calx of mercury, seemed to me suitable for the object I had in view : for nobody to-day is unaware that this substance can be reduced without addition of charcoal at a very moderate degree of heat.

In order to be sure that precipitated mercury was a true metallic calx, that it gave the usual results and the usual kind of air on reduction by the ordinary method, that is to say, using the recognized expression, by the addition of phlogiston, I mixed 1 ounce of this calx with 48 grains of powdered charcoal, and introduced the whole into a little glass retort of at most two cubic inches capacity, which I placed in a reverberatory furnace of proportionate size. The neck of this retort was about a foot long and three to four lines in diameter ; it had been bent in a flame in different places and its tip was such that it could be fixed under a bell-jar of sufficient size, filled with water, and turned upside down in a trough of water. . . .

As soon as a flame was applied to the retort and the heat had begun to take effect, the ordinary air contained in the retort expanded, and a small quantity passed into the bell-jar; but in view of the small size of the part of the retort that remained empty, this air could not introduce a sensible error, and at the most it could scarcely amount to a cubic inch. When the retort began to get hotter, air was very rapidly evolved and bubbled up through the water into the bell-jar; the operation did not last more than three-quarters of an hour, the flame being used sparingly during this interval. When the calx of mercury was reduced and air was no longer evolved, I marked the height at which the water stood in the bell-jar and found that the air set free amounted to 64 cubic inches, without allowing for the volume necessarily dissolved in the water.

I submitted this air to a large number of tests, which I will not describe in detail, and found :—

(1) that it combined with water on shaking and gave to it all the properties of acidulated, gaseous, or aerated waters such as those of Seltz, Bougues, Bussang, Pyrmont, etc. ;
(2) that animals placed in it died in a few seconds ;
(3) that candles and all burning bodies were instantly extinguished therein ;
(4) that it precipitated lime water ;
(5) that it combined very easily with fixed or volatile alkalis, removing their causticity and giving them the power of crystallizing.

All these are precisely the qualities of the kind of air known as *fixed air*, such as I obtained by the reduction of *minium* by powdered charcoal such as is set free from calcareous earths and effervescent alkalis by their combination with acids, or from fermenting vegetable matters, etc. It was thus certain that precipitated mercury gave the same products as other metallic calces on reduction in the presence of phlogiston and that it could consequently be included in the general category of metallic calces.

It remained to examine this calx alone, to reduce it without addition, to see if some elastic fluid was still set free, and if so, to determine the nature of such fluid. With this in view, I put into a retort of the same size as before (2 cubic inches) one ounce of precipitated mercury alone ; I arranged the apparatus in the same way as for the preceding experiment, so that all the circumstances were exactly the same ; the reduction was a little harder to bring about than when charcoal was present ; it required more heat and there was no perceptible effect till the retort began to get slightly red-hot ; then air was set free little by little, and passed into the bell-jar, and by keeping up the same degree of heat for $2\frac{1}{2}$ hours, all the mercury was reduced.

The operation completed, the amount of air in the bell-jar was found to be 78 cubic inches ; from this it follows, if the loss of weight of the mercury is attributed to the loss of this air, that each cubic inch must weigh a little less than two-thirds of a grain, which does not differ much from the weight of ordinary air.

Having established these results, I hastened to submit the 78 cubic inches of air I had obtained to all the tests which could indicate its nature, and I found, much to my surprise :—

(1) that it did not combine with water on shaking ;
(2) that it did not precipitate lime water, but only caused in it an almost imperceptible turbidity ;
(3) that it entered into no compounds with fixed or volatile alkalis ;
(4) that it did not in the least diminish their causticity ;
(5) that it could be used again for the calcination of metals ;
(6) in short, that it had none of the properties of fixed air : far from causing animals to perish like fixed air, it seemed on the contrary more suited to support their respiration ; not only

were candles and burning objects not extinguished, but the flame increased in a very remarkable manner : it gave much more light than in common air ; charcoal burned with a flash almost like that of phosphorus, and all combustible bodies were consumed with astonishing speed. All these circumstances fully convinced me that this air, far from being fixed air, was in a more respirable and combustible, and therefore in a purer condition, than even the air in which we live.

This seems to prove that the principle which unites with metals when they are calcined and causes them to increase in weight is nothing else than the purest part of the air which surrounds us, which we breathe, and which during calcination passes from a condition of expansibility [as a gas] to that of solidity ; if it is obtained in the form of fixed air from metallic reductions in which charcoal is employed, this is due to the combination of the charcoal with the pure part of the air, and it is very probable that all metallic calces would give, like that of mercury, only this eminently respirable air, if one could reduce them all without addition, as precipitated mercury is reduced.

In that last paragraph Lavoisier gives for the first time in history the true explanation of combustion as a more or less vigorous combination with his " pure air " (which he later called oxygen). Progress now was rapid, for Lavoisier had shown that no material substance disappeared in any chemical change, and a strictly quantitative technique could be followed. In other words, simple mathematics could be applied to chemistry, and as the mathematician would say, that is the hall-mark of any science.

Now, with a careful weighing of all the substances taking part in any chemical change, came new discoveries of immense importance. *Who laid the foundations of the atomic theory?* The answer is given in the following extracts from John Dalton's *New System of Chemical Philosophy*.

JOHN DALTON

ON THE CONSTITUTION OF BODIES

THERE are three distinctions in the kinds of bodies, or three states, which have more especially claimed the attention of philosophical chemists ; namely, those which are marked by the terms *elastic fluids, liquids, and solids.* A very famous instance is exhibited to us in water, of a body, which, in certain circumstances, is capable of assuming all the three states. In steam we recognize a perfectly elastic fluid, in water a perfect liquid, and in ice a complete solid. These observations have tacitly led to the conclusion which seems universally adopted, that all bodies of sensible magnitude, whether liquid or solid, are constituted of a vast number of extremely small particles or atoms of matter bound together by a force of attraction, which is more or less powerful according to circumstances. . . .

Whether the ultimate particles of a body, such as water, are all alike, that is, of the same figure, weight, etc., is a question of some importance. From what is known, we have no reason to apprehend a diversity in these particulars : if it does exist in water, it must equally exist in the elements constituting water, namely, hydrogen and oxygen. Now it is scarcely possible to conceive how the aggregates of dissimilar particles should be so uniformly the same. If some of the particles of water were heavier than others, if a parcel of the liquid on any occasion were constituted principally of these heavier particles, it must be supposed to affect the specific gravity of the mass, a circumstance not known. Similar observations may be made on other substances. Therefore, we may conclude that the *ultimate particles of all homogeneous bodies are perfectly alike in weight, figure, etc.* In other words, every particle of water is like every other particle of water ; every particle of hydrogen is like every other particle of hydrogen, and so on.

When any body exists in the elastic state, its ultimate particles are separated from each other to a much greater distance than in any other state ; each particle occupies the centre of a comparatively large sphere, and supports its dignity by keeping all the rest, which by their gravity, or otherwise, are disposed to encroach upon it, at a respectful distance. When we attempt to conceive the *number* of particles in an atmosphere, it is somewhat like attempting to conceive the number of stars in the universe ; we are confounded with the thought. But if we limit the subject, by taking a given volume of any gas, we seem persuaded that, let the divisions be ever so minute, the number of particles must be finite ; just as in a given space of the universe, the number of stars and planets cannot be infinite.

Chemical analysis and synthesis go no farther than to the separation of particles one from another, and to their reunion. No new creation or destruction of matter is within the reach of chemical agency. We might as well attempt to introduce a new planet into the solar system, or to annihilate one already in existence, as to create or destroy a particle of hydrogen. All the changes we can produce, consist in separating particles that are in a state of cohesion or combination, and joining those that were previously at a distance.

In all chemical investigations, it has justly been considered an important object to ascertain the relative *weights* of the simples which constitute a compound. But unfortunately the enquiry has terminated here ; whereas from the relative weights in the mass, the relative weights of the ultimate particles or atoms of the bodies might have been inferred, from which their number and weight in various other compounds would appear, in order to assist and to guide future investigations, and to correct their results. Now it is one great object of this work, to show the importance and advantage of ascertaining *the relative weights of the ultimate particles, both of simple and compound bodies, the number of simple elementary particles which constitute one compound particle, and the number of less compound particles which enter into the formation of one more compound particle.*

If there are two bodies, A and B, which are disposed to combine, the following is the order in which the combinations may take place, beginning with the most simple : namely,

1 atom of A + 1 atom of B = 1 atom of C, binary.
1 atom of A + 2 atoms of B = 1 atom of D, ternary.
2 atoms of A + 1 atom of B = 1 atom of E, ternary.
1 atom of A + 3 atoms of B = 1 atom of F, quaternary.
3 atoms of A + 1 atom of B = 1 atom of G, quaternary.
etc., etc.

[Dalton then gives some general rules which should be applied to all investigations into the building up of compounds, the guiding principle being the choice of the simplest explanation where more than one is possible.]

From the application of these rules, to the chemical facts already well ascertained, we deduce the following conclusions : 1st, That water is a binary compound of hydrogen and oxygen and the relative weights of the two elementary atoms are as 1 : 7, nearly ; 2nd, That ammonia is a binary compound of hydrogen and azote [nitrogen], and the relative weights of the two atoms are as 1 : 5, nearly ; 3rd, That nitrous gas is a binary compound of azote and oxygen, the atoms of which weigh 5 and 7 respectively. . . . That carbonic oxide is a binary compound, consisting of one atom of charcoal, and one of oxygen, together weighing nearly 12 ; that carbonic acid is a ternary compound, (but sometimes binary) consisting of one atom of charcoal, and two of oxygen, weighing 19 ; etc., etc. In all these cases the weights are expressed in atoms of hydrogen, each of which is denoted by unity.

Some of Dalton's figures were wrong ; all his ideas have had to be revised, and yet so sound were they that, on the foundations which he laid, the whole modern theory of the structure of matter has been built. Perhaps the first important addition was a realization of the differences between an atom and a molecule. While the chemists concentrated on the composition of the molecules, the physicists studied their speeds and their sizes. *What particles travelling at* 17 *miles a minute are continually hitting us ?*

For the answer we turn to Clerk Maxwell, the first Director of the Cavendish Laboratory at Cambridge. As we read it we shall also see how the heat of a body can be regarded as due to the rapid motion of its molecules : the hotter the body the higher the velocity. Thus in our study of energy as well as of matter we must go to the fundamental particles.

CLERK MAXWELL

MOLECULES

A N atom is a body which cannot be cut in two. A molecule is the smallest possible portion of a particular substance. No one has ever seen or handled a single molecule. Molecular science, therefore, is one of those branches of study which deal with things invisible, and imperceptible by our senses and which cannot be subjected to direct experiment.

A drop of water may be divided into a certain number, and no more, of portions similar to each other. Each of these the modern chemist calls a molecule of water. But it is by no means an atom, for it contains two different substances, oxygen and hydrogen, and by a certain process the molecule may be actually divided into two parts, one consisting of oxygen and the other of hydrogen.

We all know that air or any other gas placed in a vessel presses against the sides of the vessel, and against the surface of any body placed within it. On the kinetic theory this pressure is entirely due to the molecules striking against these surfaces, and thereby communicating to them a series of impulses which follow each other in such rapid succession that they produce an effect which cannot be distinguished from that of a continuous pressure.

If the velocity of the molecules is given, and the number varied, then since each molecule, on an average, strikes the sides of the vessel the same number of times, and with an impulse of the same magnitude, each will contribute an equal share to the whole pressure. The pressure in a vessel of given size is therefore proportional to the number of molecules in it, that is, to the quantity of gas in it.

This is the explanation of the fact discovered by Robert Boyle, that the pressure of air is proportional to its density. It shows also that of different portions of gas forced into a vessel, each produces its own part of the pressure independently of the rest, and this whether these portions be of the same gas or not.

Let us next suppose that the velocity of the molecules is increased. Each molecule will now strike the sides of the vessel a greater number of times in a second, but, besides this, the impulse of each blow will be increased in the same proportion, so that the part of the pressure due to each molecule will vary as the square of the velocity. Now the increase of velocity corresponds, on our theory, to a rise of temperature, and in this way we can explain the effect of warming the gas, and also the law discovered by Charles that the proportional expansion of all gases between given temperatures is the same.

The dynamical theory also tells us what will happen if molecules of different masses are allowed to knock about together. Those of greater mass will go slower than the smaller ones, so that, on an average, every molecule, great or small, will have the same energy of motion. The important consequence follows that a cubic centimetre of every gas at standard temperature and pressure contains the same number of molecules. But we must now descend to particulars, and calculate the actual velocity of a molecule of hydrogen.

A cubic centimetre of hydrogen, at the temperature of melting ice and at a pressure of one atmosphere, weighs 0·00008954 gramme. We have to find at what rate this small mass must move (whether altogether or in separate molecules makes no difference) so as to produce the observed pressure on the sides of the cubic centimetre. This is the calculation which was first made by Dr. Joule, and the result is 1859 metres per second. This is what we are accustomed to call a great velocity. It is greater than any velocity obtained in artillery practice. The velocity of other gases is less, but in all cases it is very great as compared with that of bullets.

We have now to conceive of the molecules of the air in this room flying about in all directions, at the rate of about seventeen miles in a minute.

If all these molecules were flying in the same direction, they would constitute a wind blowing at the rate of seventeen miles a minute, and the only wind which approaches this velocity is that which proceeds from the mouth of a cannon. How, then, are you and I able to stand here? Only because the molecules happen to be flying in different directions, so that those which strike against our backs enable us to support the storm which is beating against our faces. Indeed, if this molecular bombardment were to cease, even for an instant, our veins would swell, our breath would leave us, and we should, literally, expire.

But it is not only against us or against the walls of the room that the molecules are striking. Consider the immense number of them, and the fact that they are flying in every possible direction, and you will see that they cannot avoid striking each other. Every time that two molecules come into collision, the paths of both are changed, and they go off in new directions. Thus each molecule is continually getting its course altered, so that, in spite of its great velocity, it may be a long time before it reaches any great distance from the point at which it set out.

I have here a bottle containing ammonia—a gas which everyone knows by its smell. Its molecules have a velocity of 600 metres per second, so that if their course had not been interrupted by striking against the molecules of air in the room, anyone in the farthest corner would have smelt ammonia whenever I opened the bottle—before, indeed, I was able to pronounce the name of the gas. But instead of this, each molecule of ammonia is so jostled about by the molecules of air that it is sometimes going one way and sometimes another, and like a hare, which is always doubling, though it goes at a great pace, it makes very little progress. Nevertheless, the smell of ammonia is now beginning to be perceptible at some distance from the bottle. The gas does diffuse itself through the air, though the process is a slow one, and if we could close up everything in the room so as to make it air-tight, and leave everything to itself for some time, the ammonia would become uniformly mixed or inter-diffused through every part of the air in the room.

If we wish to form a mental representation of what is going on among the molecules in calm air, we cannot do better than observe a swarm of bees, when every individual bee is flying furiously, first in one direction and then in another, while the swarm, as a whole, either remains at rest, or sails slowly through the air.

The greater the velocity of the molecules and the farther they travel before their paths are altered by collision with other molecules, the more rapid will be the diffusion. Now we know already the velocity of the molecules, and therefore, by experiments on diffusion, we can determine how far, on an average, a molecule travels without striking another. This distance is called the mean path of a molecule, and it is a very small distance, quite imperceptible to us even with our best microscopes. Roughly speaking, it is about the tenth part of the length of a wave of light, which is a very small quantity. Of course the time spent on so short a path by such swift molecules must be very small. I have calculated the number of collisions which each must undergo in a second. They are reckoned by thousands of millions. No wonder that the travelling power of the swiftest molecule is but small, when its course is completely changed thousands of millions of times in a second.

The theory of liquids is not so well understood as that of gases, but the principal difference between a gas and a liquid seems to be that in a gas each molecule spends the greater part of its time in describing its free path, and is for a very small portion of its time engaged in encounters with other molecules, whereas in a liquid the molecule has hardly any free path, and is always in a state of close encounter with other molecules.

The molecules of an element keep to a constant type with a precision which is not to be found in the sensible properties of the bodies which they constitute. In the first place, the mass of each individual molecule, and all its other properties, are absolutely unalterable. In the second place, the properties of all molecules of the same kind are absolutely identical.

We can procure specimens of oxygen from very different sources—from the air, from water, from rocks of every geological epoch. The history of these specimens has been very different, and if, during thousands of years, difference of circumstances could produce difference of properties, these specimens of oxygen would show it.

In like manner we may procure hydrogen from water, from coal, or even from meteoric iron. Take two litres of any specimen of hydrogen, it will combine with exactly one litre of any specimen of oxygen, and will form exactly two litres of water.

Now if, during the whole previous history of either specimen, whether imprisoned in the rocks, flowing in the sea, or careering through unknown regions with the meteorites, any modification of the molecules had taken place, these relations would no longer be preserved.

But we have another and an entirely different method of comparing the properties of molecules. The molecule, though indestructible, is not a hard, rigid body, but is capable of internal movements, and when these are excited it emits rays, the wave-length of which is a measure of the time of vibration of the molecule.

By means of the spectroscope, the wave-lengths of different kinds of light may be compared to within one ten-thousandth part. In this way it has been ascertained, not only that

molecules taken from every specimen of hydrogen in our
laboratories have the same set of periods of vibration, but that
light, having the same set of periods of vibration, is emitted
from the sun and from the fixed stars.

We are thus assured that molecules of the same nature as
those of our hydrogen exist in those distant regions, or at least
did exist when the light by which we see them was emitted.

In the heavens we discover by their light, and by their light
alone, stars so distant from each other that no material thing
can ever have passed from one to another ; and yet this light,
which is to us the sole evidence of the existence of these distant
worlds, tells us also that each of them is built up of molecules
of the same kind as those which we find on earth. A molecule
of hydrogen, for example, whether in Sirius or Arcturus,
executes its vibrations in precisely the same time.

Natural causes, as we know, are at work which tend to
modify, if they do not at length destroy, all the arrangements
and dimensions of the earth and the whole solar system. But
though in the course of ages catastrophes have occurred and
may yet occur in the heavens, though ancient systems may be
dissolved and new systems evolved out of their ruins, the
molecules out of which these systems are built—the foundation
stones of the material universe—remain unbroken and unworn.

Maxwell mentioned the " certain process " by which the molecule
of water can be divided into its constituent parts. *Of what two gases is
water composed ?* The answer is given to us by Faraday, and since the
means employed are electrical this is only just, for it is to Faraday's
discoveries that we owe the dynamo and the electric motor and all that
they have made possible in this " electric age ".

MICHAEL FARADAY

THE DECOMPOSITION OF WATER

WE have the power of arranging the zinc which you have seen acting upon the water by the assistance of an acid, in such a manner as to cause all the power to be evolved in the place where we require it. I have behind me a voltaic pile, and I am about to show you its character and power. I hold here the extremities of the wires which transport this power from behind me, and which I shall cause to act on the water.

A great power of combustion is possessed by potassium, or zinc, or iron-filings ; but none of them show such energy as this. I will make contact between the two terminal wires of the battery : what a brilliant flash of light is produced ! This light is, in fact, produced by a forty-zinc power of burning : it is a power that I can carry about in my hands, through these wires, at pleasure—although, if I applied it wrongly to myself, it would destroy me in an instant, for it is a most intense thing, and the power you see here put forth, if I allow the spark to last while you count five, is equivalent to the power of several thunderstorms, so great is its force.

I am now going to apply this force to water to pull it to pieces, to see what else there is in the water besides hydrogen ; because if we pass steam through an iron tube, we by no means get the weight of water back which we put in, in the form of steam, though we have a very large quantity of gas evolved. We have now to see what is the other substance present. What effect has an electric current on water ? Here are two little platinum plates which I intend to make the ends of the battery, and this is a little vessel so shaped as to enable me to take it to pieces and show you its construction. In those two cups I pour mercury, which touches the ends of the wires connected with the platinum plates. In the vessel I pour some water containing a little acid (but which is put only for the purpose of facilitating the action ; it undergoes no change in the process), and connected with the top of the vessel is a bent glass tube which now passes under the jar.

I have now adjusted this apparatus, and we will proceed to affect the water some way or other. In the other case, I sent the water through a tube which was made red-hot; I am now going to pass the electricity through the contents of this vessel. Perhaps I may boil the water; if I do boil the water, I shall get steam; and you know that steam condenses when it gets cold, and you will therefore see by that whether I do boil the water or not. Perhaps, however, I shall not boil the water, but produce some other effect. You shall have the experiment and see. There is one wire which I will put to this side, and here is the other wire which I will put to the other side, and you will soon see whether any disturbance takes place. Here it is seeming to boil up famously; but does it boil? Let us see whether that which goes out is steam or not. I think you will soon see the jar will be filled with vapour, if that which rises from the water is steam. But can it be steam? Why, certainly not; because there it remains, you see, unchanged. There it is standing over the water, and it cannot therefore be steam, but must be a permanent gas of some sort. What is it? Is it hydrogen? Is it anything else? Well, we will examine it. If it is hydrogen, it will burn. I will now apply a light to it. You see it is certainly combustible, but not combustible in the way that hydrogen is. Hydrogen would not have given you that noise; but the colour of that light, when the thing did burn, was like that of hydrogen: it will, however, burn without contact with the air. That is why I have chosen this form of apparatus, for the purpose of pointing out to you what are the particular circumstances of this experiment.

In place of an open vessel I have taken one that is closed; and I am going to show you that that gas, whatever it may be, can burn without air, and in that respect differs from a candle, which cannot burn without the air. And our manner of doing this is as follows: I have here a glass vessel which is fitted with two platinum wires through which I can apply electricity ; and we can put the vessel on the air-pump and exhaust the air, and when we have taken the air out we can fasten it on to this jar, and let into the vessel that gas which was formed by the action of the voltaic battery upon the water, and which we have produced by changing the water into it—for I may go as far as this and say we have really, by that experiment, changed the water into that gas. We have not only altered its condition, but we have changed it really and truly into that gaseous substance, and all the water is there which was decomposed by the experiment. As I screw this vessel on here and make the tubes well connected, and when I open the stopcocks, if you watch the level of the water you will see that the gas will rise. I will now close the stopcocks, as I have drawn up as much as the vessel can hold, and I will pass an electric spark, from an induction coil, through the gas. The vessel was quite clear and bright at first, but it has now become dim with a deposit of water. I will again connect it to our gas reservoir, for that is what the jar really is : and as I open the stopcocks you see that the water rises : this indicates that the glass vessel must be filling. " But why is the jar empty after each explosion ? " you may ask. Because the vapour or gas into which that water has been resolved by the battery explodes under the influence of the spark, and changes into water ; and by and by you will see in this upper vessel some drops of water trickling down the sides and collecting at the bottom.

We are here dealing with water entirely, without reference to the atmosphere. The water of the candle had the atmosphere helping to produce it ; but in this way it can be produced independently of the air. Water, therefore, ought to contain that other substance which the candle takes from the air, and which, combining with the hydrogen, produces water.

I will now dip the poles—the metallic ends of this battery— into water, and see what will happen when they are kept far apart. I place one here and the other there, and I have little shelves with holes which I can put upon each pole, and so arrange them that whatever escapes from the two ends of the battery will appear as separate gases ; for you saw that the water did not become vaporous, but gaseous. The wires are now in perfect and proper connexion with the vessel containing the water ; and you see the bubbles rising ; let us collect these bubbles and see what they are. Here is a glass cylinder ; I fill it with water and put it over one end of the pile ; and I will take another and put it over the other end of the pile. And so now we have a double apparatus, with both places delivering gas. Both these jars will fill with gas. There they go, that to the right filling very rapidly ; the one to the left filling not so rapidly. I should have twice as much in this as I have in that. Both these gases are colourless ; they stand over the water without condensing ; they are alike in all things —I mean in all apparent things ; and we have an opportunity of examining these bodies and ascertaining what they are. Their bulk is large, and we can easily apply experiments to them. I will take this jar first, and will ask you to be prepared to recognize hydrogen.

Think of all its qualities—the light gas which stood well in inverted vessels, burning with a pale flame at the mouth of the jar—and see whether this gas does not satisfy all these conditions. If it be hydrogen, it will remain here while I hold this jar inverted. It burns when a light is applied to the mouth of the jar ; it is evidently hydrogen.

What is there now in the other jar ? You know that the two together made an explosive mixture. But what can this be which we find as the other constituent in water, and which must therefore be that substance which made the hydrogen burn ? We know that the water we put into the vessel consisted of the two things together. We find one of these in hydrogen : what must that other be which was in the water before the experiment, and which we now have by itself? I am about to put this lighted splinter of wood into the gas. The gas itself will not burn, but it will rekindle the glowing splinter. See how it invigorates the combustion of the wood, and how it makes it burn far better than the air would make it burn; and now you see by itself that very other substance which is contained in the water, and which, when the water was formed by the burning of the candle, must have been taken from the atmosphere. What shall we call it, A, B, or C? Let us call it O—call it " oxygen " : it is a very good, distinct-sounding name. This, then, is the oxygen which was present in the water, forming so large a part of it.

We shall now begin to understand more clearly our experiments and researches ; because, when we have examined these things once or twice, we shall soon see why a candle burns in the air. When we have in this way analysed the water—that is to say, separated, or electrolysed its parts out of it—we get two volumes of hydrogen, and one of the body that burns it. And these two are represented to us on the following diagram, with their weights also stated ; and we shall find that the oxygen is a very heavy body by comparison with the hydrogen. It is the other element in water.

Oxygen	.	.	88·9
Hydrogen	.	.	11·1
			———
Water	.	.	100·0

It was experiments such as these which enabled a more accurate figure to be given to all atomic weights. You will observe that the weight of oxygen is almost exactly eight times that of the weight of hydrogen, and since the volume of hydrogen is twice that of oxygen we have an atomic weight of 16 for oxygen. Other atomic weights were also found to be whole numbers, or nearly so, and in spite of the fact that some weights obstinately refused to approach any whole number the idea that there might be some primordial brick out of which all the atoms were built began to gain ground once again. As early as 1815 Prout suggested that this " brick " was hydrogen. The theory gained support from the fact that the elements could be placed in groups, showing a regular increase in atomic weights and a strong family likeness in each group. This idea was worked out by Mendeléef in his Periodic Classification in 1869. He went so far as to prophesy that certain missing places in his table would be filled by newly discovered elements, and these predictions were quickly justified.

A new line of investigations was opened out when Sir J. J. Thomson at the Cavendish Laboratory discovered the electron, and measured its charge and its mass. Here was a smaller, and possibly even more fundamental " brick " than the atom of hydrogen. Further secrets were revealed by the discovery of an " exhibitionist " atom, that of radium. *How did the discovery of radium solve many of the problems connected with atomic structure ?*

SIR J. J. THOMSON

RADIUM AND RADIO-ACTIVITY

A STRIKING discovery like that of the Röntgen rays acts much like the discovery of gold in a sparsely populated country ; it attracts workers who come in the first place for the gold, but who may find that the country has other products, other charms, perhaps, even more valuable than the gold itself. The country in which the gold was discovered in the case of the Röntgen rays was the department of physics dealing with the discharge of electricity through gases, a subject which, almost from the beginning of electrical science, had attracted a few enthusiastic workers, who felt convinced that the key to unlock the secret of electricity was to be found in a vacuum tube. Röntgen, in 1895, showed that when electricity passed through such a tube, the tube emitted rays which could pass through bodies opaque to ordinary light ; which could, for example, pass through the flesh of the body and throw a shadow of the bones on a suitable screen. The fascination of this discovery attracted many workers to the subject of the discharge of electricity through gases, and led to great improvements in the instruments used in this type of research. It is not, however, to the power of probing dark places, important though this is, that the influence of Röntgen rays on the progress of science has mainly been due; it is rather because these rays make gases, and, indeed, solids and liquids, through which they pass conductors of electricity. It is true that before the discovery of these rays other methods of making gases conductors were known, but none of these were so convenient for the purpose of accurate measurement.

The study of gases exposed to Röntgen rays has revealed in such gases the presence of particles charged with electricity; some of these particles are charged with positive, others with negative electricity.

The properties of these particles have been investigated; we know the charge they carry, the speed with which they move under an electric force, the rate at which the oppositely charged ones re-combine, and these investigations have thrown a new light not only on electricity, but also on the structure of matter.

Radio-activity was brought to light by the Röntgen rays. One of the many remarkable properties of these rays is to excite phosphorescence in certain substances, including the salts of uranium, when they fall upon them. Since Röntgen rays produce phosphorescence, it occurred to Becquerel to try whether phosphorescence would produce Röntgen rays. He took some uranium salts which had been made to phosphoresce by exposure, not to Röntgen rays but to sunlight, tested them, and found that they gave out rays possessing properties similar to Röntgen rays. Further investigations showed, however, that to get these rays it was not necessary to make the uranium phosphoresce, that the salts were just as active if they had been kept in the dark. It thus appeared that the property was due to the metal and not to the phosphorescence, and that uranium and its compounds possessed the power of giving out rays which, like Röntgen rays, affect a photographic plate, make certain minerals phosphoresce, and make gases through which they pass conductors of electricity.

Niepce de Saint-Victor had observed some years before this

discovery that paper soaked in a solution of uranium nitrate affected a photographic plate, but the observation excited little interest. The ground had not then been prepared, by the discovery of the Röntgen rays, for its reception, and it withered and was soon forgotten.

Shortly after Becquerel's discovery of uranium, Schmidt found that thorium possessed similar properties. Then Monsieur and Madame Curie, after a most difficult and laborious investigation, discovered two new substances, radium and polonium, possessing this property to an enormously greater extent than either thorium or uranium, and this was followed by the discovery of actinium by Debierne. Now the researches of Rutherford and others have led to the discovery of so many new radio-active substances that any attempt at christening seems to have been abandoned, and they are denoted, like policemen, by the letters of the alphabet.

Mr. Campbell has recently found that potassium, though far inferior in this respect to any of the substances I have named, emits an appreciable amount of radiation, the amount depending only on the quantity of potassium, and being the same whatever the source from which the potassium is obtained or whatever the elements with which it may be in combination.

The radiations emitted by these substances are of three types, known as α, β, and γ rays. The α rays have been shown by Rutherford to be positively electrified atoms of helium, moving with speeds which reach up to about one-tenth of the velocity of light. The β rays are negatively electrified corpuscles, moving in some cases with very nearly the velocity of light itself, while the γ rays are unelectrified, and are analogous to the Röntgen rays.

The radio-activity of uranium was shown by Crookes to arise from something mixed with the uranium, and which differed sufficiently in properties from uranium itself to enable it to be separated by chemical analysis. He took some uranium, and by chemical treatment separated it into two portions, one of which was radio-active and the other not.

Next Becquerel found that if these two portions were kept for several months, the part which was not radio-active to begin with gained radio-activity, while the part which was radio-active to begin with had lost its radio-activity. These effects and many others receive a complete explanation by the theory of radio-active change which we owe to Rutherford and Soddy.

According to this theory, the radio-active elements are not permanent, but are gradually breaking up into elements of lower atomic weight; uranium, for example, is slowly breaking up, one of the products being radium, while radium breaks up into a radio-active gas called radium emanation, the emanation into another radio-active substance, and so on, and that the radiations are a kind of swan song emitted by the atoms when they pass from one form to another; that, for example, it is when a radium atom breaks up and an atom of the emanation appears that the rays which constitute the radio-activity are produced.

Thus, on this view of the atoms the radio-active elements are not immortal; they perish after a life whose average value ranges from thousands of millions of years in the case of uranium to a second or so in the case of the gaseous emanation from actinium.

When the atoms pass from one state to another they give out large stores of energy; thus their descendants do not inherit the whole of their wealth of stored-up energy, the estate becomes less and less wealthy with each generation; we find, in fact, that the politician when he imposes death duties is but imitating a process which has been going on for ages in the case of these radio-active substances.

Many points of interest arise when we consider the rate at which the atoms of radio-active substances disappear. Rutherford has shown that whatever be the age of these atoms, the percentage of atoms which disappear in one second is always the same; another way of putting it is that the expectation of life of the atom is independent of its age—that an atom of radium a thousand years old is just as likely to live for another thousand years as one just sprung into existence.

Now this would be the case if the death of the atom were due to something from outside which struck old and young indiscriminately ; in battle, for example, the chance of being shot is the same for old and young ; so that we are inclined at first to look to something coming from outside as the cause why an atom of radium, for example, suddenly changes into an atom of the emanation. But here we are met with the difficulty that no changes in the external conditions that we have as yet been able to produce have had any effect on the life of the atom ; as far as we know at present the life of a radium atom is the same at the temperature of a furnace as at that of liquid air—it is not altered by surrounding the radium by thick screens of lead or other dense materials to ward off radiation from outside, and, what to my mind is especially significant, it is the same when the radium is in the most concentrated form, when its atoms are exposed to the vigorous bombardment from the rays given off by the neighbouring atoms, as when it is in the most dilute solution, when the rays are absorbed by the water which separates one atom from another. This last result seems to me to make it somewhat improbable that we shall be able to split up the atoms of the non-radio-active elements by exposing them to the radiation from radium ; if this radiation is unable to effect the unstable radio-active atoms, it is somewhat unlikely that it will be able to affect the much more stable non-radio-active elements.

The energy developed by radio-active substances is exceedingly large, one gramme of radium developing nearly as much energy as would be produced by burning a ton of coal. This energy is mainly in the a particles, the positively charged helium atoms which are emitted when the change in the atom takes place ; if this energy were produced by electrical forces it would indicate that the helium atom had moved through a potential difference of about two million volts on its way out of the atom of radium. The source of this energy is a problem of the deepest interest; if it arises from the repulsion of similarly electrified systems exerting forces varying inversely as the square of the distance, then to get the requisite amount of energy the systems, if their charges were comparable with the charge on the a particle, could not, when they start, be further apart than the radius of a corpuscle, 10^{-13} cm. If we suppose that the particles do not acquire this energy at the explosion, but that before they are shot out of the radium atom they move in circles inside this atom with the speed with which they emerge, the forces required to prevent particles moving with this velocity from flying off at a tangent are so great that infinite charges of electricity could only produce them at distances comparable with the radius of a corpuscle.

The properties of radium have consequences of enormous importance to the geologist as well as to the physicist or chemist. In fact, the discovery of these properties has entirely altered the aspect of one of the most interesting geological problems, that of the age of the earth. Before the discovery of radium it was supposed that the supplies of heat furnished by chemical changes going on in the earth were quite insignificant, and that there was nothing to replace the heat which flows from the hot interior of the earth to the colder crust. Now when the earth first solidified it only possessed a certain amount of capital in the form of heat, and if it is continually spending this capital and not gaining any fresh heat it is evident that the process cannot have been going on for more than a certain number of years, otherwise the earth would be colder than it is. Lord Kelvin, in this way, estimated the age of the earth to be less than 100 million years. Though the quantity of radium in the earth is an exceedingly small fraction of the mass of the earth, only amounting, according to the determinations of Professors Strutt and Joly, to about five grammes in a cube whose side is 100 miles, yet the amount of heat given out by this small quantity of radium is so great that it is more than enough to replace the heat which flows from the inside to the outside of the earth. This, as Rutherford has pointed out, entirely vitiates the previous method of determining the age of the earth. The fact is that the radium gives out so much heat that we do not quite know what to do with it, for if there was as much radium throughout the interior of the earth as there is in its crust, the temperature of the earth would increase much more rapidly than it does as we descend below the earth's surface. Professor Strutt has shown that if radium behaves in the interior of the earth as it does at the surface, rocks similar to those in the earth's crust cannot extend to a depth of more than forty-five miles below the surface.

It is remarkable that Professor Milne from the study of earthquake phenomena had previously come to the conclusion that rocks similar to those at the earth's surface only descend a short distance below the surface ; he estimates this distance at about thirty miles, and concludes that at a depth greater than this the earth is fairly homogeneous.

Though the discovery of radio-activity has taken away one method of calculating the age of the earth, it has supplied another.

The gas helium is given out by radio-active bodies, and since, except in beryls, it is not found in minerals which do not contain radio-active elements, it is probable that all the helium in these minerals has come from these elements. In the case of a mineral containing uranium, the parent of radium, in radio-active equilibrium with radium and its products, helium will be produced at a definite rate. Helium, however, unlike the radio-active elements, is permanent and accumulates in the mineral ; hence if we measure the amount of helium in a sample of rock and the amount produced by the sample in one year we can find the length of time the helium has been accumulating, and hence the age of the rock. This method, which is due to Professor Strutt, may lead to determinations not merely of the average age of the crust of the earth but of the ages of particular rocks and the date at which the various strata were deposited ; he has, for example, shown in this way that a specimen of the mineral thorianite must be more than 240 million years old.

It was Thomson's experiments on the discharge of electricity through rarefied gases which led to the discovery of the electron, and electrical methods have since been used in all further research on the atom. *Why must this be so?* Thomson's lecture on the Atomic Theory which is here given will suggest the answer, and at the same time give us a summary of the knowledge gained by 1914.

SIR J. J. THOMSON

THE ATOMIC THEORY

IT was not until 1803, the date of Dalton's atomic theory, that the conception of the atom played any considerable part in scientific discovery. Dalton's theory was based on the proportions by weight of the different elements in various chemical compounds ; he showed that these proportions are exactly those which would exist if each element consisted of a great number of particles, all the particles of any one element being exactly alike, but each element having its own particular kind of particle.

For some time after Dalton's enunciation of his theory, no very important advances were made in our knowledge of atoms, but in the second half of the nineteenth century the atomic theory was greatly advanced by the work of Clausius, Clerk Maxwell, Boltzmann, Joule, Kelvin, and Willard Gibbs on the kinetic theory of gases. These philosophers showed that many of the properties of gases can be explained if the gas is regarded as a collection of a very large number of small particles in rapid motion. Some important results as to the size of atoms were obtained in this way, but the greater part of the researches of this time had to do with the properties of swarms of atoms, and threw but little light on the constitution of the individual atom.

In fact, it was not until quite the close of the nineteenth century, when attention was turned to the study of electrified atoms instead of unelectrified ones, that our acquaintance with the atom became at all intimate.

The advance made through the electrification of the atom has been most remarkable ; it is due to the fact that an un-electrified atom is so elusive that unless more than a million million are present we have no means sufficiently sensitive to detect them. Or, to put it in another way, unless we had a better test for a man than we have for an unelectrified molecule, we should be unable to find out that the earth was inhabited.

The electrified atom or molecule, on the other hand, is much more assertive, so much so that it has been found possible in some cases to detect the presence of a single electrified atom. A billion unelectrified atoms may escape our observation, whereas a dozen or so electrified ones are detected without difficulty.

One reason why electrified atoms and molecules are so much easier to study is that we can subject them to forces far more intense than any we can apply to unelectrified ones ; we can exert much more control over them, and force them into situations where their habits can be observed.

For example, if a mixture of different kinds of electrified atoms is moving along in one stream, then, when electric and magnetic forces are applied to the stream simultaneously, the different kinds of atoms are sorted out, and the original stream is divided up into a number of smaller streams separated from each other. The particles in any one of the smaller streams are all of the same kind.

Thus, if the original stream contained a mixture of hydrogen and oxygen atoms, it would, by the action of the electric and magnetic forces, be split up into two separate streams, one con-sisting exclusively of oxygen atoms, the other of hydrogen atoms. We shall call the streams into which the original stream is split up the electric spectrum of the atoms. By means of it we can analyse a stream of atoms, just as a beam of light is analysed by sending it through a spectroscope and observing the different rays into which it is divided.

By means of the electric spectrum we can prove in a very direct and striking way some of the fundamental truths of the atomic theory—that a gas contains only a few kinds of particles, that all the particles of one kind have exactly the same mass, and there are molecules as well as atoms in the majority of gases, that some gases, such as helium and mercury vapour, where there is only one stream instead of two, have atoms but no molecules.

But when we analyse in this way a gas through which an electric discharge is passing, we find, along with the atoms and molecules, particles of an altogether different type. These particles are always charged with negative electricity, and their mass is an exceedingly small fraction, $\frac{1}{1800}$, of that of the smallest atom known, the atom of hydrogen. They are so small that their volume bears to that of the atom much the same proportion as that between a small pellet and a large room.

These particles are called electrons, or corpuscles, and no matter what the nature of the gas may be, whether it is hydrogen, helium, or mercury vapour, the electrons or corpuscles remain unchanged in quality. In fact, there is only one kind of electron, and we can get it out of every kind of matter. The conclusion is irresistible that the electron or corpuscle is a constituent of every atom, and that we are able, by forces that we have even now at our command, to detach it from the atom.

Since the electron can be got from all the chemical elements, we may conclude that electrons are a constituent of all atoms. By ingenious devices we are even able to calculate the number of electrons in the atoms. The number is not very far from half the atomic weight; thus in the carbon atom there would be six electrons, in the oxygen atom eight, and so on, while in the lightest atom, hydrogen, there is probably only one. This is a most interesting result when we remember that there is room for 1800 of these corpuscles in an atom of hydrogen.

The constant difference between the number of electrons in the atom of one element and that in the atom of the element next in the series is strong evidence in favour of the view that the atoms of the consecutive elements differ from each other by the addition of a primordial atom, which apparently is the atom of helium.

It is probable that the electrons in an atom, if they exceed a certain number, are divided up into groups, into a series of spherical layers, like the coatings of an onion, separated from each other by finite distances, the number of such layers depending upon the number of electrons in the atom, and thus upon its atomic weight.

The electrons in the outside layer will be held in their places less firmly than those in the inner layers ; they are more mobile, and will arrange themselves more easily under the forces exerted upon them by other atoms. On the outside belt, therefore, depend what we may call the social qualities of the atom in relation to other atoms ; while the electrons in the strata nearer the centre of the atom, which are much more firmly held, have to do with the more intrinsic properties of the atom.

The several ways of investigating the structure of the atom all involve great labour, and anyone who has used them must often have felt what a boon it would have been if we had an eye which would enable us to have a good look at an atom and have done with it. Now I cannot say that any such eye has been invented, but Mr. C. T. R. Wilson has made some approach to it by a beautiful method by which we can see, not indeed the individual atom itself, but still the path of such an atom, and in some cases what is going on in the atom.

The method is based on the principle that when charged atoms or electrons are produced in air sufficiently super-saturated with water vapour, the water condenses on them and nowhere else. Thus each atom or electron is surrounded by a little drop of water, and the regions where they are produced are mapped out by threads of little drops of water resembling seed pearls ; these can be photographed and studied at leisure. Now an electrified atom or electron travelling through a gas when it strikes against the atoms knocks some of the electrons out of them, and thus leaves behind it a trail of electrified wrecks. Mr. Wilson deposits drops of water on these wrecks, and thus the path of the electrified atom or electron is marked by a trail of drops of water which can be seen and photographed. We can map out in this way the path of even one atom.

Though what we know about the atom is but a minute fraction of what there is to know, some very important conclusions about atoms have been established on what seems strong evidence in the course of the last few years. We know, for example, that there are such things as atoms, that the atoms of an element are all of one kind, that atoms of different elements contain a common constituent—the corpuscle or electron, about which we know a good deal. We know, too, the number of electrons in an atom.

We have strong evidence that the electrons in the atom are divided into groups, and that some properties of the atom, those which we associate with the innermost group, are connected in a very simple way with the total number of electrons in the atom ; that there are other properties, notably the chemical ones, which change in a rhythmical way with the atomic weight of the element, and which depend upon the electrons near the surface of the atom.

Lastly, we know that there are regions in the atom, probably the most interesting of all, about which we know little or nothing, whose investigation will provide intensely interesting work for many generations of physicists, who will most assuredly have no reason to be " mournful that no new wonder may betide ".

Electrons certainly form part of all atoms, and electrons are units of negative electricity. *Is all matter composed of electricity?* Certainly it seems to be very largely if not entirely so, as we shall see from this lecture by Lord Rutherford, possibly the greatest of the explorers who mapped out for us the incredibly small country of the atom. It is taken from his Presidential Address to the British Association in 1923.

LORD RUTHERFORD

THE ELECTRICAL STRUCTURE OF MATTER

IT is my intention this evening to refer very briefly to some of the main features of that great advance in knowledge of the nature of electricity and matter which is one of the salient features of the interval since the last meeting of this Association in Liverpool.[1]

In order to view the extensive territory which has been conquered by science in this interval, it is desirable to give a brief summary of the state of knowledge of the constitution of matter at the beginning of this epoch. Ever since its announcement by Dalton the atomic theory has steadily gained ground, and formed the philosophic basis for the explanation of the facts of chemical combination. In the early stages of its application to physics and chemistry it was unnecessary to have any detailed knowledge of the dimensions or structure of the atom. It was only necessary to assume that the atoms acted as individual units, and to know the relative masses of the atoms of the different elements. In the next stage, for example, in the kinetic theory of gases, it was possible to explain the main properties of gases by supposing that the atoms of the gas acted as minute perfectly elastic spheres. During this period, by the application of a variety of methods, many of which were due to Lord Kelvin, rough estimates had been obtained of the absolute dimensions and mass of the atoms. These brought

[1] In 1896.

out the minute size and mass of the atom and the enormous number of atoms necessary to produce a detectable effect in any kind of measurement. From this arose the general idea that the atomic theory must of necessity for ever remain unverifiable by direct experiment, and for this reason it was suggested by one school of thought that the atomic theory should be banished from the teaching of chemistry, and that the law of multiple proportions should be accepted as the ultimate fact of chemistry.

While the vaguest ideas were held as to the possible structure of atoms, there was a general belief among the more philosophically minded that the atoms of the elements could not be regarded as simple unconnected units. The periodic variations of the properties of the elements brought out by Mendeléef were only explicable if atoms were similar structures in some way constructed of similar material. We shall see that the problem of the constitution of atoms is intimately connected with our conception of the nature of electricity. The wonderful success of the electro-magnetic theory had concentrated attention on the medium or ether surrounding the conductor of electricity, and little attention had been paid to the actual carriers of the electric current itself. At the same time the idea was generally gaining ground that an explanation of the results of Faraday's experiments on electrolysis was only possible on the assumption that electricity, like matter, was atomic in nature. The name " electron " had even been given to this fundamental unit by Johnstone Stoney, and its magnitude roughly estimated, but the full recognition of the significance and importance of this conception belongs to the new epoch.

For the clarifying of these somewhat vague ideas, the proof in 1897 of the independent existence of the electron as a mobile electrified unit, of mass minute compared with that of the lightest atom, was of extraordinary importance. It was soon seen that the electron must be a constituent of all the atoms of matter, and that optical spectra had their origin in their vibrations. The discovery of the electron and the proof of its liberation by a variety of methods from all the atoms of matter was of the utmost significance, for it strengthened the view that the electron was probably the common unit in the structure of atoms which the periodic variation of the chemical properties had indicated. It gave for the first time some hope of the success of an attack on that most fundamental of all problems—the detailed structure of the atom. In the early development of this subject science owes much to the work of Sir J. J. Thomson, both for the boldness of his ideas and for his ingenuity in developing methods for estimating the number of electrons in the atom, and of probing its structure. He early took the view that the atom must be an electrical structure, held together by electrical forces, and showed in a general way lines of possible explanation of the variation of physical and chemical properties of the elements, exemplified in the periodic law.

In the meantime our whole conception of the atom and of the magnitude of the forces which held it together was revolutionized by the study of radio-activity. The discovery of radium was a great step in advance, for it provided the experimenter with powerful sources of radiation specially suitable for examining the nature of the characteristic radiations which are emitted by the radio-active bodies in general. It was soon shown that the atoms of radio-active matter were undergoing spontaneous transformation, and that the characteristic radiations emitted, viz. the a, β, and γ rays, were an accompaniment and consequence of these atomic explosions. The wonderful succession of changes that occur in uranium, more than thirty in number, was soon disclosed and simply interpreted on the transformation theory. The radio-active elements provide us for the first time with a glimpse into Nature's laboratory, and allow us to watch and study but not control the changes that have their origin in the heart of the radio-active atoms.

These atomic explosions involve energies which are gigantic

compared with those involved in any ordinary physical or chemical process. In the majority of cases an a particle is expelled at high speed, but in others a swift electron is ejected, often accompanied by a γ ray, which is a very penetrating X-ray of high frequency. The proof that the a particle is a charged helium atom for the first time disclosed the importance of helium as one of the units in the structure of the radioactive atoms, and probably also in that of the atoms of most of the ordinary elements. Not only, then, have the radioactive elements had the greatest direct influence on natural philosophy, but in subsidiary ways they have provided us with experimental methods of almost equal importance. The use of a particles as projectiles with which to explore the interior of the atom has definitely exhibited its nuclear structure, has led to artificial disintegration of certain light atoms, and promises to yield more information yet as to the actual structure of the nucleus itself.

The influence of radio-activity has also extended to yet another field of study of fascinating interest. We have seen that the first rough estimates of the size and mass of the atom gave little hope that we could detect the effect of a single atom. The discovery that the radio-active bodies expel actual charged atoms of helium with enormous energy altered this aspect of the problem. The energy associated with a single a particle is so great that it can readily be detected by a variety of methods. Each a particle, as Sir Wm. Crookes first showed, produces a flash of light easily visible in a dark room when it falls on a screen coated with crystals of zinc sulphide. This scintillation method of counting individual particles has proved invaluable in many researches, for it gives us a method of unequalled delicacy for studying the effects of single atoms. The a particle can also be detected electrically or photographically, but the most powerful and beautiful of all methods is that perfected by Mr. C. T. R. Wilson for observing the track through a gas not only of an a particle but of any type of penetrating radiation which produces ions or electrified particles along its path. The method is comparatively simple, depending on the fact, first discovered by him, that if a gas saturated with moisture is suddenly cooled, each of the ions produced by the radiation becomes the nucleus of a visible drop of water. The water-drops along the track of the a particle are clearly visible to the eye, and can be recorded photographically. These beautiful photographs of the effect produced by single atoms or single electrons appeal, I think, greatly to all scientific men. They not only afford convincing evidence of the discrete nature of these particles, but give us new courage and confidence that the scientific methods of experiment and deduction are to be relied upon in this field of inquiry; for many of the essential points brought out so clearly and concretely in these photographs were correctly deduced long before such confirmatory photographs were available. At the same time, a minute study of the detail disclosed in these photographs gives us most valuable information and new clues on many recondite effects produced by the passage through matter of these flying projectiles and penetrating radiations.

In the meantime a number of new methods had been devised to fix with some accuracy the mass of the individual atom and the number in any given quantity of matter. The concordant results obtained by widely different physical principles gave great confidence in the correctness of the atomic idea of matter. The method found capable of most accuracy depends on the definite proof of the atomic nature of electricity and the exact valuation of this fundamental unit of charge. We have seen that it was early surmised that electricity was atomic in nature. This view was confirmed and extended by a study of the charges carried by electrons, a particles, and the ions produced in gases by X-rays and the rays from radio-active matter. It was first shown by Townsend that the positive or negative charge carried by an ion in gases was invariably equal to the charge carried by the hydrogen ion in the electrolysis of water, which we have seen was assumed,

and assumed correctly, by Johnstone Stoney to be the fundamental unit of charge. Various methods were devised to
measure the magnitude of this fundamental unit; the best
known and most accurate is Millikan's, which depends on
comparing the pull of an electric field on a charged droplet
of oil or mercury with the weight of the drop. His experiments
gave a most convincing proof of the correctness of the electronic
theory, and gave a measure of this unit, the most fundamental
of all physical units, with an accuracy of about one in a
thousand. Knowing this value, we can by the aid of electro-
chemical data easily deduce the mass of the individual atoms
and the number of molecules in a cubic centimetre of any gas
with an accuracy of possibly one in a thousand, but certainly
better than one in a hundred. When we consider the minute-
ness of the unit of electricity and of the mass of the atom, this
experimental achievement is one of the most notable even in
an era of great advances.

The idea of the atomic nature of electricity is very closely
connected with the attack on the problem of the structure of
the atom. If the atom is an electrical structure it can only
contain an integral number of charged units, and, since it is
ordinarily neutral, the number of units of positive charge
must equal the number of negative. . . .

It may be of interest to try to visualize the conception of
the atom we have so far reached by taking for illustration the
heaviest atom, uranium. At the centre of the atom is a minute
nucleus surrounded by a swirling group of ninety-two electrons, all in motion in definite orbits, and occupying but by
no means filling a volume very large compared with that of
the nucleus. Some of the electrons describe nearly circular
orbits round the nucleus; others, orbits of a more elliptical
shape whose axes rotate rapidly round the nucleus. The
motion of the electrons in the different groups is not necessarily confined to a definite region of the atom, but the
electrons of one group may penetrate deeply into the region
mainly occupied by another group, thus giving a type of

interconnexion or coupling between the various groups. The maximum speed of any electron depends on the closeness of the approach to the nucleus, but the outermost electron will have a minimum speed of more than 1000 kilometres per second, while the innermost K electrons have an average speed of more than 150,000 kilometres per second, or half the speed of light. When we visualize the extraordinary complexity of the electronic system we may be surprised that it has been possible to find any order in the apparent medley of motions.

In reaching these conclusions, which we owe largely to Professor Bohr and his co-workers, every available kind of data about the different atoms has been taken into consideration. A study of the X-ray spectra, in particular, affords information of great value as to the arrangement of the various groups in the atom, while the optical spectrum and general chemical properties are of great importance in deciding the arrangements of the superficial electrons. While the solution of the grouping of the electrons proposed by Bohr has been assisted by considerations of this kind, it is not empirical in character, but has been largely based on general theoretical considerations of the orbits of electrons that are physically possible on the generalized quantum theory. The real problem involved may be illustrated in the following way. Suppose the gold nucleus be in some way stripped of its attendant seventy-nine electrons and that the atom is reconstituted by the successive addition of electrons one by one. According to Bohr, the atom will be reorganized in one way only, and one group after another will successively form and be filled up in the manner outlined. The nucleus atom has often been likened to a solar system where the sun corresponds to the nucleus and the planets to the electrons. The analogy, however, must not be pressed too far. Suppose, for example, we imagined that some large and swift celestial visitor traverses and escapes from our solar system without any catastrophe to itself or the planets. There will inevitably result permanent changes in the lengths of the month and year, and our system will never return to its original state. Contrast this with the effect of shooting an electron or a particle through the electronic structure of the atom. The motion of many of the electrons will be disturbed by its passage, and in special cases an electron may be removed from its orbit and hurled out of its atomic system. In a short time another electron will fall into the vacant place from one of the outer groups, and this vacant place in turn will be filled up, and so on until the atom is again reorganized. In all cases the final state of the electronic system is the same as in the beginning. This illustration also serves to indicate the origin of the X-rays excited in the atom, for these arise in the process of reformation of an atom from which an electron has been ejected, and the radiation of highest frequency arises when the electron is removed from the K group. . . .

I must now bring to an end my survey, I am afraid all too brief and inadequate, of this great period of advance in physical science. In the short time at my disposal it has been impossible for me, even if I had the knowledge, to refer to the great advances made during the period under consideration in all branches of pure and applied science. I am well aware that in some departments the progress made may justly compare with that of my own subject. In these great additions to our knowledge of the structure of matter every civilized nation has taken an active part, but we may be justly proud that this country has made many fundamental contributions. With this country I must properly include our Dominions overseas, for they have not been behindhand in their contributions to this new knowledge. It is, I am sure, a matter of pride to this country that the scientific men of our Dominions have been responsible for some of the most fundamental discoveries of this epoch, particularly in radio-activity.

This tide of advance was continuous from 1896, but there was an inevitable slackening during the War. It is a matter of good omen that, in the last few years, the old rate of progress has not only been maintained but even intensified, and there appears to be no obvious sign that this period of great advances has come to an end. There has never been a time when the enthusiasm of the scientific workers was greater, or when there was a more hopeful feeling that great advances were imminent. This feeling is no doubt in part due to the great improvement during this epoch of the technical methods of attack, for problems that at one time seemed unattackable are now seen to be likely to fall before the new methods. In the main, the epoch under consideration has been an age of experiment, where the experimenter has been the pioneer in the attack on new problems. At the same time, it has been also an age of bold ideas in theory, as the quantum theory and the theory of relativity so well illustrate.

In watching the rapidity of this tide of advance in physics I have become more and more impressed by the power of the scientific method of extending our knowledge of Nature. Experiment, directed by the disciplined imagination either of an individual, or still better, of a group of individuals of varied mental outlook, is able to achieve results which far transcend the imagination alone of the greatest natural philosopher. Experiment without imagination, or imagination without recourse to experiment, can accomplish little, but, for effective progress, a happy blend of these two powers is necessary. The unknown appears as a dense mist before the eyes of men. In penetrating this obscurity we cannot invoke the aid of supermen, but must depend on the combined efforts of a number of adequately trained ordinary men of scientific imagination. Each in his own special field of inquiry is enabled by the scientific method to penetrate a short distance, and his work reacts upon and influences the whole body of other workers. From time to time there arises an illuminating conception, based on accumulated knowledge, which lights up a large region and shows the connexion between these individual efforts, so that a general advance follows. The attack begins anew on a wider front, and often with improved technical weapons. The conception which led to this advance often appears simple and obvious when once it has been put forward. This is a common experience, and the scientific man often feels a sense of disappointment that he himself had not foreseen a development which ultimately seems so clear and inevitable.

The intellectual interest due to the rapid growth of science to-day cannot fail to act as a stimulus to young men to join in scientific investigation. In every branch of science there are numerous problems of fundamental interest and importance which await solution. We may confidently predict an accelerated rate of progress of scientific discovery, beneficial to mankind certainly in a material but possibly even more so in an intellectual sense. In order to obtain the best results certain conditions must, however, be fulfilled. It is necessary that our universities and other specific institutions should be liberally supported, so as not only to be in a position to train adequately young investigators of promise, but also to serve themselves as active centres of research. At the same time there must be a reasonable competence for those who have shown a capacity for original investigation. Not least, peace throughout the civilized world is as important for rapid scientific development as for general commercial prosperity. Indeed, science is truly international, and for progress in many directions the co-operation of nations is as essential as the co-operation of individuals. Science, no less than industry, desires a stability not yet achieved in world conditions.

There is an error far too prevalent to-day that science progresses by the demolition of former well-established theories. Such is very rarely the case. For example, it is often stated that Einstein's general theory of relativity has overthrown the work of Newton on gravitation. No statement could be farther from the truth. Their works, in fact, are hardly comparable, for they deal with different fields of thought. So far as the work of Einstein is relevant to that of Newton, it is simply a generalization and broadening of its basis ; in fact, a typical case of mathematical and physical development. In general, a great principle is not discarded but so modified that it rests on a broader and more stable basis.

It is clear that the splendid period of scientific activity which we have reviewed to-night owes much of its success and intellectual appeal to the labours of those great men in the past who wisely laid the sure foundations on which the scientific worker builds to-day, or to quote from the words inscribed in the dome of the National Gallery: " The works of those who have stood the test of ages have a claim to that respect and veneration to which no modern can pretend ".

Lord Rutherford stresses the fact that scientific principles are not discarded altogether as the result of further progress, but are more often put on a firmer and broader basis. A striking example is given in the modifications made to Dalton's atomic theory. The main plank in this theory was that all atoms of the same substance had the same mass. Sir J. J. Thomson took the electrified atoms of a rarefied gas, passed them through electric and magnetic fields, and showed that, since they then produced a definite line on a photographic plate and not a blur, they must all have at least approximately the same mass. In this way the first experimental proof of Dalton's century-old theory was given. But only a year or two later Dr. Aston (also of the Cavendish Laboratory) with improved apparatus and a special form of camera, found that the neon gas, so widely used in glow-advertisements, instead of giving a single line gave a definitely double one. Thus no sooner was Dalton's theory given experimental proof than a striking exception was found. Here was a challenge which Dr. Aston at once accepted. It led to the discovery that most of the elements have atoms of two or more slightly different weights, and it is the mixture of these different atoms— " isotopes " as they are called—which accounts for the previously awkward fact that many atomic weights are not whole numbers. *What may well be described as the most beautiful and delicate series of experiments in modern science ?* They were made by Dr. Aston, and are briefly and modestly mentioned in the following extracts from an address to the Röntgen Society.

F. W. ASTON

THE STRUCTURE OF THE ATOM

WHEN the Council asked me to accept this high office, in addition to the honour of adding my name to so distinguished a company as the past presidents of the Röntgen Society, there were two other considerations which influenced my decision. One was that Röntgen's great discovery proved to be a dominant factor in my own life, for it was the main event which led me to take up the profession of research worker in physics. The second was that although now, for many years, my work has been related but distantly to that of radiology, it gives me pleasure to think that I was one of the first in this country to experiment with the X-rays, and if actual manufacture of the apparatus for their production counts, I can claim a seniority exceeded by few of my audience to-night. I may therefore be excused if I commence my address with a few personal reminiscences. As one who has struggled with apparatus which was capable of going wrong in every conceivable way—and several others—I am sure of a sympathetic hearing from an audience of radiologists, especially the older ones.

The announcement of the discovery of X-rays affected me profoundly, so much so that, although it is thirty years ago, I can remember, as if it was yesterday, reading the short paragraph in the *Birmingham Daily Post* which described some of their amazing properties. Being a poor mathematician I had advanced very little in physics, but knew enough from Professor Poynting to make me profoundly doubtful of these road-hogging radiations which drove through a conductor, like aluminium, in entire disregard of Maxwell's white line. Although on this particular occasion my sceptical attitude of mind proved needless, I have not yet entirely abandoned it. A little caution in accepting sensational reports is by no means out of place.

My acquaintance with the rays commenced at Mason

College, Birmingham, where, aided by the equipment of the Physics Department, we were able to make conversazione demonstrations of their uncanny screen effects soon after the discovery was made known. During what may be called the " paleoradiological " era, when the wall of the bulb formed the anticathode, the fact that glass could be heated to its melting point by a stream of cathode rays could be demonstrated in a very striking manner. My actual personal investigations started at my old home, in a hayloft which was subsequently converted into a laboratory. Purchase of apparatus was quite beyond my means, so that I had to make everything literally from scratch. I blew my own bulbs and the Sprengel pump to exhaust them ; very good for the development of one's patience is the use of the Sprengel pump ; I wound my own coil—eight weary miles of wire—by hand, and stole the family jam-pots to make the bichromate batteries to run it. The inefficiency of the coil and the debility of the cells—I suppose they only gave about 10 watts—pleased me little enough at the time, but I have good reason to be thankful for them to-day. Had really powerful sources been available I must long ago have lost my hands, as did so many of my less fortunate contemporaries.

The phenomena attending the slow exhaustion of a discharge tube are among the most beautiful in the whole field of physics, and I suppose they will never be more fascinating than at that time when so little was known, so much was possible. " Age cannot wither, custom cannot stale, their infinite variety ", and I can still take a childish delight in watching the preliminary exhaustion of the discharge tube of my mass-spectrograph, so expeditiously carried out by modern methods —that is, supposing there is no indication of a leak.

I have here the most successful, and almost the first, of my early X-ray focus tubes. It dates about 1896, and was blown from an ordinary chemical test-tube. The leads and the anticathode were made from the platinum wire and foil supplied at that time in all student's chemical sets. The current passed through it must have been very minute, for, as far as I can recollect, a hand took about a quarter of an hour's exposure. As regards definition of small bones in structures such as birds, mice, or lizards, it was quite surprisingly good. With larger and living creatures it was not so successful, and the ludicrous results of an attempt to screen our pet pug's tail I will leave in obscurity. There is nothing sacred to the enthusiastic investigator. Of what use is the second commandment to the radiologist ? He will make a graven image in the likeness of *anything* that is in the heavens above, or the earth beneath, or the water that is under the earth.

During the following years, although engaged in other work, I did occasional experiments with my apparatus, and in 1903, having observed that there were conditions under which the Crookes' dark space was accurately measurable, I returned to Mason College to start my first serious physical research on that fascinating phenomenon. In 1911, continuing this line of work at the Cavendish Laboratory, I was able to show, by a beautiful method devised by Sir J. J. Thomson, that the distribution of potential in the dark space was such that the electric force at any point varied linearly from zero in the negative glow to a maximum at the surface of the cathode. . . .

After ten years' continuous work on the dark space I was led, by the influence of Sir J. J. Thomson, to study more especially these positive ions which, if the cathode is pierced, pass through the aperture and constitute the positive rays. The analysis of positive rays has led to considerable advances in our knowledge of the nature of matter, so, with the transfer of my inquiries from the front to the back of the cathode, I will turn from autobiographical remarks to my more serious subject.

THE STRUCTURE OF THE ATOM

The idea that all atoms of matter might be built of the same primordial units, that is to say, might differ not in material but only in construction, dates back at least as far as Prout. This philosopher endeavoured over a century ago to show that atoms of all elements were themselves built of atoms of hydrogen. A little earlier Dalton had postulated, in probably the most important theory in the whole history of chemistry, that atoms of the same element were of equal weight. If both these theories were right, the atomic weights of all elements would be comparable with each other as whole numbers. This the chemists soon found was quite incompatible with experimental evidence. They had to choose between the two theories, and chose the one that was untrue. In this they were perfectly right, for it is more important that a scientific theory should be simple than that it should be true.

The point cannot be tested by chemical methods, for these require a vast number of atoms, and so can only yield a mean result. The way in which Dalton's postulate was first attacked and shown to be incorrect was in the province of radioactivity, when Soddy showed that lead which was produced from thorium minerals had a different atomic weight to the lead which was produced from uranium minerals. This meant that substances could exist which had identical chemical properties but different atomic weights ; these Soddy called isotopes. This reasoning could not be applied to ordinary elements. For these there is only one conclusive test, which is to compare the weights of individual atoms. It is here that positive rays are of such value, for they are atoms carrying a positive charge and moving with so high a speed that they can be detected by a fluorescent screen or photographic plate.

The first experimental comparison of the weights of individual atoms was made by Sir J. J. Thomson by his " parabola " method, in which the rays are subjected to electric and magnetic fields giving deflections at right angles to each other. Subjected to this test, many of the elements seemed to obey Dalton's rule, giving single or apparently single parabolic streaks expected from groups of atoms travelling with different velocities but all of the same mass. But results obtained with neon suggested that in this gas the atoms were of two different weights, 20 and 22, the accepted atomic weight being 20·20. The accuracy of the parabola method of analysis was not sufficient to prove the point, but this was done by means of the mass spectrograph.   I gave an account of this instrument to you five years ago, so it will be enough for me to recall that in it, by using electric and magnetic fields giving deflections at 180° to each other, it is possible to focus the rays and obtain a spectrum dependent on mass alone.   By measurements of this mass-spectrum it is possible to compare the weights of atoms to 1 part in 1000.   In this way a satisfactory proof was obtained that neon did consist of two isotopes, 20 and 22, which, present in the proportion of 9 to 1, give the mean atomic weight 20·2. Chlorine, whose chemical atomic weight is 35·46, was found to consist of two isotopes, 35 and 37.   Many of the elements, such as carbon, oxygen, nitrogen, etc., were found to be " simple ", that is, to consist of atoms all of the same weight, but even more were found to be " complex ", mixtures of two or more isotopes.   Selenium, krypton, cadmium, and mercury each have six, tin probably eight, and xenon possibly nine isotopic constituents.   In all, fifty-six out of the eighty known non-radioactive elements have been analysed into their constituent isotopes or shown to be simple.[1]

By far the most important result of these measurements is that, with the exception of hydrogen, the weights of the atoms of all the elements measured, and therefore almost certainly of all elements, are whole numbers to the accuracy of experiment, namely, about 1 part in 1000.   Of course, the error expressed in fractions of a unit increases with the weight measured, but with the light elements the divergence from whole numbers is extremely small.   This generalization, which is called the *whole number rule*, has removed the only serious obstacle to the electrical theory of matter.   It enables us to restate Prout's original hypothesis with the modification that the primordial

[1] At the present time all the common elements have been analysed and some 260 isotopes discovered. The discoveries of the neutron and other fundamental particles have altered ideas on the structure of the nucleus in detail, but the main principles remain the same.

atoms are of two kinds : protons and electrons, the atoms of positive and negative electricity. The proton is very much smaller and heavier than the electron, actually about 1850 times as heavy. According to the nucleus atom theory, which we owe to Sir Ernest Rutherford, all the protons and about half the electrons are packed very close together to form a central positively charged nucleus, round which the remaining electrons circulate, somewhat like planets round a sun. All the spectroscopic and chemical properties of the atom depend on the positive charge on the nucleus, which is the excess of protons over electrons. This is clearly the number of planetary electrons in the neutral atom ; it is called the *atomic number* and is actually the number of the element in the periodic classification : 1 for Hydrogen, 2 for Helium, 3 for Lithium, and so on. The whole-number weight of the atom, on the other hand, will be the total number of neutral pairs of protons and electrons it contains. This is also the number of protons in its nucleus, and is called the *mass-number* of the atom : 1 for Hydrogen, 4 for Helium, 6 and 7 for the isotopes of Lithium, and so on. Atoms are isotopic, that is, belong to the same element, when their nuclei have the same net positive charge, but they may have a different total number of protons, and so different weights.

We picture the atom as consisting of a central nucleus and an outer system of electrons, but when we come to inquire into the dimensions of the electrical particles themselves in relation to the dimensions of the atoms they compose, we are faced with a very surprising result. The protons and electrons are infinitesimal compared with the atom. To convey any direct idea of the numerical relations is almost hopeless, and were we to construct a scale model of the atom as big as the dome of St. Paul's, we should have some difficulty in seeing the electrons, which would be little larger than pin-heads, while the protons in the nucleus would escape notice altogether as dust particles invisible to the naked eye. If we represent the nucleus of a helium atom as the size of a pea, its planetary electrons would be about a quarter of a mile away. Experimental evidence leaves us no escape from the conclusion that matter is empty. An atom, even of so heavy an element as lead, is as empty as the solar system and only occupies the spherical space we allot to it by virtue of the rapid and continuous rotation of its outer electrons. Led by the knowledge that under certain conditions these outer electrons could be stripped from the atom, and the nuclei thereby enabled to approach closer to each other, Eddington was able to predict that in certain stars matter could attain a density thousands of times greater than the greatest we know. As no doubt you are all aware, this prediction has been strikingly verified by recent observations on the companion of Sirius, which at the same time have afforded another signal triumph for Einstein's relativity theory.

We have heard a good deal of loose talk in recent years of " splitting " the atom. Whenever you draw your fountain pen from your pocket you split countless millions of atoms in the sense that you violently tear planetary electrons away from them, by the friction between the ebonite and the cloth. This form of splitting is called ionization. In it the atom suffers no sort of permanent injury. It simply captures the first electron it can to replace the one it has lost, and after notifying the world at large of its recovery by a wireless signal it goes on exactly as before. In such a solid as copper the exchange of electrons from one atom to another can be effected with the greatest freedom, and it is the passage of these loose electrons which constitutes the ordinary electric current. I suspect that the high conductivity of the negative glow, and of flames is due to an exchange of a somewhat similar kind.

I mentioned the despatch of a wireless signal sent by the atom on repair of its injury. The type of this radiation depends on the extent of the damage done. For superficial effects it is light and radiant heat, for deeper and more violent ones it is X-rays. The displacement of the innermost and most tightly bound electrons gives rise to the hardest X-rays. The tightness of binding depends on the nuclear charge, so that for the emission and absorption of the hardest rays the heaviest elements must be employed. It is to be emphasized that in all such cases we are only concerned with the outer electrons. With the nucleus it is a very different state of affairs. To dislodge any part of this requires violence of an altogether higher order, but if it is done the whole atom is changed, and changed permanently. This is no longer ionization but transmutation.

TRANSMUTATION OF THE ELEMENTS

Transmutation of the elements, so long sought by the alchemists, takes place spontaneously in the radioactive atoms whose nuclei are unstable and periodically eject helium nuclei and electrons, which are the well-known alpha and beta rays. Several claims of artificial transmutation of elements have been made recently in serious scientific journals. I will deal with the more doubtful ones first. Three years ago it was stated that helium was formed when a tungsten wire was deflagrated by an intense discharge. Sir Ernest Rutherford pointed out the extreme improbability of any disruption of the tungsten nucleus under these conditions, and a careful repetition of the experiments, with greater precautions, proved that he was right. Quite recently in this country a claim has been made that helium has been produced by transmutation in a vacuum tube discharge. If true, this would be the greatest discovery in history, but the detection at the same time of neon, another atmospheric gas, is, to my mind, a very suspicious circumstance, and when these alchemists seriously suggest that success or failure may depend on the use of a particular form of obsolete make and break, my sceptical nose is aware of a whiff of stuffed alligator.

A much more interesting case is that of the liberation of gold from mercury by electric discharge, even in an ordinary mercury vapour lamp. Here similar experimental results have been obtained by several investigators in different parts of the world, and the quantities of gold produced remarkably large—large enough as we shall see to dissipate the hope, so confidently expressed, that it is formed by transmutation of the mercury atoms owing to the addition of an electron to their nuclei. This claim was supported, I confess much to my surprise, by a well-known authority, on the ground that since the nucleus is positively charged it would be quite easy to fire an electron into it. This is pushing the analogy of the sun and planet system to unjustifiable length. We know that a planet directed towards the sun would actually fall into it, but if every time an electron was directed towards a nucleus it fell into it and was absorbed, how could matter have a permanent existence at all ? We know there must be some mechanism in Nature which prevents such a collapse taking place in this simple manner. But even if we grant the theoretical possibility there are still fatal practical objections. The addition of an electron to the nucleus of one of the isotopes of mercury will turn it into an atom of gold, but cannot alter its weight appreciably. Now the atomic weight of the so-called artificial gold has been determined by Honigschmid, and agrees within experimental error with the value 197·2 assigned to ordinary gold. Quite recently, by means of a new and more powerful mass-spectrograph, I have been able to resolve the isotopes of mercury, and so determine its composition which was previously in some doubt. I find that it consists of 198, 199, 200, 201, 202, and 204. There is no isotope 197 previously suspected.[1] This fact, combined with the atomic weight, makes it quite certain that no transmutation of the kind claimed could produce the gold found. This is ordinary gold which must have been present in the mercury from the start. I understand that this view has now been shown to be right by the failure of the experiment when sufficient care is taken to eliminate all traces of gold from the mercury beforehand.

Unless our views on the structure of nuclei are very wide of

---

[1] This isotope has been since discovered, but is so extremely rare, 1 part in 10,000 of mercury, that the argument is not affected.

the mark, failure in such experiments is inevitable, for the forces employed are ludicrously inadequate to cause disruption. The work of Rutherford, Chadwick, Ellis, and others, leaves no doubt that just as the dimensions of the nucleus are almost inconceivably small—the radius of that of aluminium is probably less then 0·0000000000004 cm.—so the forces binding together its component parts are gigantic and to be measured in millions of volts. Such forces, I need hardly remind you, are not yet available in the laboratory.[1] They are, however, provided, on an atomic scale, in the form of the alpha particles shot out of radioactive atoms, and with these Rutherford has succeeded in producing real and definite transmutation. The method consists in bombarding the atoms with the swiftest alpha particles, which are helium nuclei with a velocity of over 100,000 miles per second, which corresponds to an energy of many millions of volts. In order to effect a disintegration, these projectiles must make a direct hit on the nucleus. When this happens in the case of most elements lighter than potassium, a proton is dislodged from the nucleus which is thereby transmuted into another element.

These observations have recently been strikingly confirmed by Blackett, who, using the beautiful Wilson fog-track method, has actually succeeded in photographing the disintegration of nitrogen nuclei struck by swift alpha particles. As I have already pointed out, the dimensions of the nucleus are minute compared with those of the atom. It can be calculated that an alpha particle colliding with an atom will only hit the nucleus once in about ten thousand million collisions, so that although each alpha particle makes about 200,000 collisions in completing its track, a very large number of photographs had to be taken. Actually some 400,000 tracks were photographed and eight disintegrations detected. In these the thin track of the dislodged proton could be clearly seen, and a somewhat unexpected feature brought out is that in each case the projectile is retained by the target. The nitrogen nucleus loses one

[1] But see page 196.

proton but captures the helium nucleus fired at it, and so would appear to become an isotope of oxygen of atomic weight 17. No such body is known in Nature,[1] which suggests that the atom so formed is not permanently stable.

ATOMIC ENERGY

In the possibility of artificial transmutation lies the hope of one day releasing the so-called " atomic " energy. The whole-number rule is not mathematically exact, and it has been shown by direct measurements on the mass-spectrograph that an atom of helium, which consists of four protons, two nuclear electrons and two planetary electrons, weighs nearly 1 per cent. less than four atoms of hydrogen, each of which consists of one proton and one electron. The number of particles is identical and the change of mass is ascribed to the different way they are arranged, and is called the packing effect. The theory of relativity tells us that mass and energy are interchangeable, and that if a mass m is destroyed a quantity of energy equal to mc^2 is produced, where c is the velocity of light. Hence, if we could transmute hydrogen into helium we should produce energy in quantities which, for any sensible amount of matter, are prodigious beyond the dreams of scientific fiction. For one gram atom of hydrogen, that is the quantity in 9 c.c. of water, the energy is :—

$$0.0077 \times 9 \times 10^{20} = 6.93 \times 10^{18} \text{ ergs.}$$

Expressed in terms of heat, this is 1.66×10^{11} calories, or in terms of work 200,000 kilowatt-hours. In a tumbler of water lies enough power to drive the *Mauretania* across the Atlantic and back at full speed. Here we have at last a supply sufficient even for the demands of astronomers, indeed there is now little doubt that the vast supply of energy radiated by the stars can be kept up for centuries by the loss of an insignificant fraction of their mass.

[1] It has since been detected by optical methods and found to occur to the extent of about 1 in 3000 of ordinary oxygen.

The search for the ultimate bricks which Nature uses has proved fascinating but tantalizing. Each step forward has made the scientists congratulate themselves on their progress, only to find that the ultimate secret was still beyond their grasp. Each hill surmounted served only to reveal another and a higher one beyond. When the electron, the unit of negative electricity had its charge measured and its mass weighed it looked as though we did at last know all about one of Nature's simplest bricks. But it was not so simple. *What is the electron ? Is it matter or energy ?* The answer—both—is given in the following extract from Sir J. J. Thomson's famous lecture *Beyond the Electron*.

SIR J. J. THOMSON

BEYOND THE ELECTRON

NOT so very long ago the atom was thought to be a terminus beyond which it was impossible, from the nature of things, to penetrate. The atom was regarded as indivisible, impenetrable, eternal, unaffected by heat, electricity or any other physical agent. The inside of the atom was regarded as a territory which the physicist could never enter. Then there came a time when the sanctuary of the atom was invaded, and it was found that the atom was built up of smaller parts—of electrons carrying a charge of negative, and of protons carrying a charge of positive electricity. Means were devised for counting the number of electrons in an atom, and it was found that the atom, instead of being just the little hard solid particle of the original view, was a very complex thing, comparable in complexity with the solar system. It was found, moreover, that it was this complexity, this fine structure inside the atom, which endowed matter with its electrical and chemical properties. If we have any insight into these properties, it is due not so much to the idea that matter consists of a large number of small particles as to the knowledge we have obtained of individual particles and their electronic structure.

Experiments told us what was the mass of the electron and the total charge of electricity associated with it; they did not, however, tell us anything about its structure; they did not tell us, for example, whether it was just a point charge of negative electricity or whether, like the atom, it was made up of smaller parts, sub-electrons and sub-protons, as it were. It is true that there is no evidence that there are different kinds of electrons as there are different kinds of atoms, but this may be because one kind of electron is so much more stable than any other that the number of the latter is quite insignificant. In the absence of definite knowledge it was natural to begin with the simplest assumption and regard the electron as a single point charge surrounded by a structureless medium. The mathematics are simpler on this view than on any other; this, however, is not conclusive, as there is no evidence that the convenience of mathematicians has been a dominant factor in the scheme of the universe.

It was therefore not improbable that, in the light of further knowledge, this view of the electron might prove as untenable as the corresponding view for the atom. My object, this afternoon, is to point out that this further knowledge has come, and that the electron and its surroundings must have a structure very different from that first assigned to them.

Perhaps some of you may ask, " Is not going beyond the electron really going too far? Ought one not to draw the line somewhere?" It is the charm of physics that there are no hard and fast boundaries, that each discovery is not a terminus but an avenue leading to country as yet unexplored, and that, however long the science may exist, there will still be an abundance of unsolved problems and no danger of unemployment for physicists.

The reason why we have to give up the old view of the electron is that it has recently been shown that a moving electron, even a uniformly moving one, is always accompanied by a series of waves. These waves, as it were, carry it along and determine the way it is to go; thus a moving electron is a much more complicated thing than a small point charge of electricity in uniform motion.

The clearest evidence for the existence of these waves round the electron is, I think, given by a research by my son, Professor G. P. Thomson, on the effects to be observed when electrons pass through exceedingly thin plates of metal—the metal has to be exceedingly thin, far thinner than the thinnest gold leaf. These plates, however, when they are obtained are exceedingly valuable physical instruments, for they enable us to test whether anything passing through them is a stream of particles or a train of waves. For suppose that we have a thin pencil of rays and we wish to determine whether these are a swarm of particles all moving in one direction or a train of waves. In either case if the pencil fell directly on a photographic plate it would produce a sharply defined image. Now let us see what would be the effect on this image of interposing the thin plate of metal. If the pencil consists of particles, these will strike against the molecules of the plate and be deflected by the collisions—how much each particle is deflected is, within certain limits, very much a matter of chance, so that some particles will be deflected more than others. Thus when they come out of the plate the particles will not all be moving in the same direction ; the stream of particles will spread out into a cone ; this will make the image they form on the photographic plate bigger and blurred, and it will become just a smudge without any definite pattern.

Suppose, however, that instead of a stream of particles we have a train of waves, then in consequence of the regular spacing of the molecules in the plate, the plate will act like a diffraction grating ; and if the distance between the molecules is comparable with the length of the waves, we know from the properties of such gratings that when we interpose the plate in the way of the beam the original spot will not become just a smudge, but will be surrounded by a series of bright rings whose radii bear definite ratios to each other.

The effect found by my son when he passed a stream of

electrons through the plate was a well-developed series of rings, and they are just in the position of the diffraction rings which would be produced if light of suitable wave length passed through the thin plate. That those rings marked the path of electrons was shown by bringing a magnet near the photographic plate ; the rings were displaced just as the path of the electrons would be displaced ; this shows that the blackening of the plate is due to electrons and not to waves of light, for these would not have been affected by the magnet. Thus it appears that the electrons in their path through the metal are bent, not like particles would be bent, but in all respects like waves of a suitable wave length. Hence we conclude that the electron is accompanied by a train of waves, and that these waves have complete control over its path ; the electron is compelled to follow the lead of the waves.

The thin metal film does more than detect the waves, it enables us to measure their wave length. My son did this, and the results are most interesting, for it turns out that these electronic waves are of extraordinary high pitch ; the pitch of the lowest is nearly a million times that of visible light, it is far higher than that of Röntgen rays, higher than all but the very hardest of the highest pitched rays hitherto known, the γ-rays emitted by radio-active substances. They introduce us to a new type of radiation whose properties may differ fundamentally from any type of radiation with which we are acquainted.

Just as it was found necessary to supplement the corpuscles which, on the old corpuscular theory constituted light, by systems of waves, so it turns out that bare corpuscles of electricity are insufficient to explain the properties of electrons, and that these, like the corpuscles of light, must be accompanied by systems of waves. This duality of corpuscles and waves seems to be in evidence in many regions of physics and may be of the nature of things.

The geography of the outlying electrons in the atoms of different elements was mapped out fairly quickly, but the central nucleus proved a much harder nut to crack, and it was only with the invention of powerful apparatus such as the cyclotron—which gradually increases the velocity of charged particles until it approaches that of light—that the nucleus could be attacked and made to give up some of its secrets. We have still only touched the fringe of this problem of Nature's ultimate bricks. Some of them have a lifetime of only two-millionths of a second! *What is the most transient form of Matter?* The answer is given in a talk by Professor C. F. Powell of Bristol University.

C. F. POWELL

THE MOST TRANSIENT FORM OF MATTER

DURING the past three years, there have been important advances in our knowledge of new types of fundamental particles which are referred to as the mesons. The name meson is derived from the fact that the masses of the particles are intermediate in value between that of the proton and that of the electron, two of the familiar bricks of which the atomic physicist regards the world as built up. The first mesons were observed in the cosmic radiation by Anderson in 1938, but recently another type has been discovered. The rapid rate of progress in the past three years, the continual interplay of theory and practice, in which the flood of new ideas put forward to explain a new world of experience has continually to be checked and corrected by the results of experiment, has provided us with one of the most interesting and exciting periods in the history of physics. I propose here to give an account of the properties of the new particles and of the methods by which they were discovered.

For those who are not specialists in physics, let me begin by outlining very briefly the main features of our present picture of the structure of matter. We now know that all the tangible things in the world around us are made up of myriads of small particles—the atoms. Each atom can be pictured as a miniature solar system in which the main mass is concentrated in a very small central core which carries a positive electric charge. According to the magnitude of this positive charge, so the neutral atom will contain different numbers of electrons. Hydrogen atoms have one electron, helium two, lithium three ; and so on through the Periodic Table of the chemist up to the heavier elements such as uranium with 92, and plutonium with 94.

With the development of our knowledge, it was found that, contrary to the views of the physicists of the last century, the atoms of matter are not indestructible. The discovery of radio-activity by the Curies showed, for example, that some of the heavy elements are unstable, that they are liable to decompose spontaneously. This radio-activity is due to the instability of the central nuclei of these atoms. In one mode of radio-active decay, a nucleus throws out a part of its own structure, which Rutherford showed to be identical with the nuclei of the atoms of the second element of the Periodic Table, namely, helium. The fact of radio-active decay suggests that, like the atoms, the nuclei must also be regarded as having a structure ; and we know that they are built up of two types of particles of nearly equal mass, neutrons and protons. The protons are identical with the nuclei of the simplest atoms, those of ordinary hydrogen, and carry a positive charge ; on the other hand, the neutrons, as their name implies, have no charge. It is sometimes convenient to use the word " nuclcon " as a general term embracing both neutrons and protons.

Although we are fairly certain about the constituents of the nuclei, we know very little about the arrangement of the nucleons in the nucleus ; and of the nature of the forces that bind them together and make most of the nuclei stable structures. Our knowledge is still rudimentary, and contrasts greatly with our technical ability to liberate energy from nuclear sources by means of the fission process.

How, then, can our knowledge be increased ? In studying

the properties of the nuclei we must solve two main technical problems. First, to get sources of fast particles with which to bombard the nuclei. Matter can only be studied by changing it, and the atomic nuclei are protected against intrusion by the screen of electrons with which they are surrounded. Further, the positive charges which they carry tend to prevent any positively charged particles from approaching them because of the electrostatic repulsion between like charges. Two of the projectiles that we commonly use—protons and alpha particles—can only enter a nucleus if they are moving with great speed : if, as the physicists say, they have great kinetic energy. In the early days of the subject, before about 1930, the only available sources of fast alpha-particles were those provided by the radio-active decay of heavy nuclei, but we now have machines producing a stream of artificially accelerated particles enormously more intense than those from radio-active substances.

But for the particles of the highest energy we still have to turn to natural sources. Coming from out of space, and incident on the earth's atmosphere, there is a thin stream of particles, mostly protons, moving with energies hundreds of times greater than those we can produce in the laboratory. Although they are so few in numbers, these cosmic-ray particles produce effects which cannot be reproduced artificially, and they enable us to study processes which are of the greatest importance for the development of nuclear physics. The particles composing the cosmic ray stream come from all directions of space, and when they enter the earth's atmosphere they begin to collide with the nuclei of the atoms of the air and break them up. The nuclei are literally knocked to pieces, though it is only rarely that the disintegration is complete in the sense that the protons and neutrons which compose a nucleus are completely scattered. Usually some of the nucleons appear bound together ; as an alpha-particle for example, which is itself made up of two neutrons and two protons, or sometimes in even larger associations.

This is where mesons come in. Some types of mesons are created in these explosive disintegrations, and such processes are occurring all the time at sea level, and even at depths underground. We find, however, that the frequency of occurrence increases rapidly with height, so it is a great advantage to be able to make experiments at high altitudes. In passing through the atmosphere, the primary particles lose energy because of collisions with nuclei, and they are eventually brought to rest. Only a small fraction of them—those which arrive at the top of the atmosphere with the greatest energy and which happen, by chance, to make only a few collisions—can penetrate to great depths. So, for cosmic-ray experiments, a number of laboratories have been established at mountain altitudes, such as that on the Pic du Midi in the French Pyrenees, the Jungfraujoch high altitude research station in Switzerland, and the station on Mount Alagez in the Caucasus, all of which are at altitudes of about 10,000 feet.

For experiments at even greater heights we can employ high flying aircraft or free balloons. To give you a rough idea of the advantages of going to these great altitudes, I may say that approximately the same number of events can be recorded in a given apparatus in the following intervals of time at different heights ; four months at sea level is equivalent to four weeks on the Jungfraujoch, to ten hours at 35,000 feet, and to one hour at 70,000 feet. This comparison is in some ways misleading, however, because some processes only occur near the top of the atmosphere. Thus the incoming protons, which make up the greater part of the incident stream, are accompanied by the nuclei of even heavier atoms in much smaller numbers. We have a rough idea of the chemical constitution of the matter of the universe from the study of astrophysics. It seems that the different types of nuclei in the cosmic-ray stream are present in the same proportion as in the matter of the universe. But the heavier nuclei lose energy in the atmosphere more rapidly than the protons because of their greater charge, and very few of them penetrate to altitudes

less than 70,000 feet, about thirteen miles high.   Experiments
with high-flying balloons are therefore of particular interest in
giving information which cannot be obtained lower down.

The second technical problem we have to face is that of
observing the effects produced by the fast particles when they
strike nuclei.   For this purpose, three instruments have been
of the greatest importance : the Geiger counter ; the Wilson
expansion chamber ;   and the photographic plate.   The
Geiger counter is essentially an electrical trigger which records
the instant of passage through the instrument of a charged
particle.   The Wilson chamber enables us to display and
photograph the tracks produced by charged particles when they
pass through the gas with which the chamber is filled.

The photographic method is, in principle, one of the oldest
in nuclear physics, but it is only in the last ten years that it
has become possible to make it precise.   This was achieved
as a result of university research using ordinary photographic
plates.   Recently, however, plates with greatly improved
characteristics have been produced, and for this development,
physicists all over the world are indebted to the skill and persistence of the British scientists in the photographic industry,
notably to Mr. Waller and Dr. Harrison, and more recently
to Dr. Berriman and Dr. Davies, who, in the last few months,
have succeeded in making plates which are sensitive to every
kind of electrically charged particle, including electrons.   The
photographic plate gives similar information to that obtained
with the Wilson chamber.   It consists of a flat sheet of glass on
which has been deposited a gelatine emulsion containing
myriads of small grains of silver bromide.   When a charged
particle passes through the emulsion, a tiny speck of silver, a
so-called latent image, is produced in some of the grains of
silver bromide lying in its path.   On developing the plate,
these particular grains are changed into grains of amorphous
silver.   The plate is subsequently " fixed ", to remove the
unchanged silver bromide, and then examined under the
microscope.   The tracks of any charged particles then appear

as a line of black grains like irregular black beads on an invisible thread. The method has the advantage of extreme simplicity, and the small weight of the plates and the absence of any auxiliary equipment make it particularly suitable for experiments with balloons. Our knowledge of mesons has been accumulated as a result of all these methods.

The original discovery of one type of mesons by Anderson was made with a Wilson chamber operated at sea level. Later, it was shown by means of Geiger counters that these mesons are extremely unstable and have an average lifetime of only two-millionths of a second, of two microseconds. The particles can be charged either positively or negatively, and they transform spontaneously into an electron and two neutral particles which are emitted from the parent particle with great kinetic energy. The accurate determination of the mass of these mesons has been made only recently and has been found to be 216 times that of the electron.

At the time of their discovery, and for some years afterwards, it was thought that the existence of the mesons provided a remarkable confirmation of certain theoretical speculations by the Japanese physicist Yukawa. Yukawa argued in the following way : When a fast electron collides with another electron, electro-magnetic forces come into play between the two particles, because of their electric charge. As a result of the collision a quantum of radiation is created : particles of the same kind of stuff as that with which we are familiar, because, when they enter our eyes, they give us the sensation of sight. By analogy Yukawa suggested that when two fast nucleons collide, the special nuclear forces responsible for the cohesion of nuclei would be brought into play; and that, as a result, a new kind of particle would be created, which he referred to as a " heavy quantum ". Now the force between two nucleons, unlike gravitational or electrical forces, only acts when the distance between the particles is very small, and from a general argument which I cannot here reproduce, he concluded that unlike the particles of ordinary radiation, the heavy quanta should have a definite mass. He thought that they should be about 300 times as heavy as an electron, that both positive and negatively charged kinds should exist, and that they should be very unstable, and decay in about a tenth of a micro-second after the instant of their creation with the emission of an electron. Some of the observed properties of the mesons discovered by Anderson were remarkably close to these predictions and this suggested that they were to be identified with Yukawa's heavy quanta. Later on, however, serious difficulties arose. The mesons cannot be arriving at the top of the atmosphere from out of space; they are too unstable, and would decay during their journey. They must, therefore, originate in the earth's atmosphere. We find, however, that they are never or only rarely created, in the process visualized by Yukawa, in collisions between nucleons. Where, then, do they come from?

The difficulties were resolved by the discovery in Bristol in 1947 of a different type of meson, the pi-meson, by means of the photographic method. We use the name pi-meson, Greek P for primary, because of certain characteristics of the particles that I shall describe. A pi-meson has a mass 286 times that of the electron. It is one of the most transient forms of matter hitherto discovered, and on the average lives for only about a hundredth of a micro-second. It can carry either a positive or a negative charge, and the two kinds can be distinguished in the following way. When the particles pass into a solid substance, they lose speed and can be brought to rest. If this solid substance is the emulsion of a photographic plate we are able to see, after the development of the plate, what happens to these particles. It is found that the positive pi-mesons are brought to rest and then give out a second meson. This particle, which we refer to as a mu-meson, Greek M, has a mass 216 times that of the electron, and it has recently been shown to be identical with the particles discovered by Anderson. It is found that the mu-meson when produced by the spontaneous decay of a pi-meson which is at rest, is always shot out at the same speed. This proves that it is accompanied by the emission of a single neutral particle which does not leave a track, and so is unobserved, but which we must assume to recoil from the mu-meson like a shot from a gun.

The negative pi-particles have a different fate. When brought to rest they are drawn towards the nucleus of an atom of the material as a result of the electrostatic attraction between unlike charges. They disappear into the nucleon and the energy corresponding to their mass is given to the nucleons so that the nucleus breaks up. In this respect they have properties similar to the particles predicted by Yukawa.

The behaviour of the positive and negative pi-particles which I have described is only observed when the particles are slowed down in moving through solid substances. When moving in the atmosphere they lose energy much less rapidly, and then, because of their short lifetime, both the positive and negative particles usually decay, while in motion. It is the pi-particles which are produced when the primary cosmic-ray particles collide with nuclei. The downward moving stream of pi-particles thus formed rapidly transforms into a stream of mu-mesons, some of which are able to penetrate down to the surface of the earth, and even to considerable depths underground. The origin of the particles discovered by Anderson is thus made clear. In the collisions of the protons with nuclei the pi-particles are produced first, and the mu-mesons appear as a product of their decay. . . .

Fifty years ago the discovery of the electrons and of the radio-activity of the heavy nuclei led to the discovery of the structure of the atoms and to the great technical discoveries for which our age will be remembered. It seems possible that the mesons will play an equally important rôle in the new world of the nucleus which we are just entering.

As our readers may have gathered, Professor Powell was himself the moving spirit in the work which led to the discovery of the two different types of meson. Our reading of his lecture will have given us a new conception of the " stability " of matter. This old world of ours with its " everlasting hills " is made up of particles which seem to be eternal, and which for the most part do approach that state of durability. But there is nothing intrinsically everlasting in them. They are made up of smaller particles which, as we have seen, can disappear as " matter " and reappear as energy. Every instant some of the particles—a very small number indeed out of the whole uncountable host—do have this eventful history. The great, the overwhelming majority do not. And that is why the ordinary " laws of nature " apply in our everyday world, even if they cannot be applied to the ultimate particles. These laws are " statistical laws ". They tell us what will happen in all reasonable probability. But they are not absolute. The nineteenth-century physicists and chemists were fond of talking about their " exact " sciences, and they were rather inclined to look down on sciences like psychology or sociology because of the incalculable reactions of human nature. They no longer do this. Professor Andrade has already drawn a parallel, in his Introduction, between psychology as applied (a) to the herd, and (b) to the atom.

The following address by Dr. Schrödinger, famous for his work on the Quantum Theory, will give us a modern answer to the question *What are the relations between matter and energy ?* and will reinforce the thought, already expressed, that when we know all about atoms we shall know all about both physics and chemistry.

ERWIN SCHRÖDINGER

MATTER DEMATERIALIZED

IF a pack of playing cards, after being thoroughly shuffled, were found in the order C.-9 ; S.-2, J ; H.-5 ; C.-A, and so on (some particular order, chosen at random) this would astonish nobody. If, however, it exhibited the order H.-A, K, Q, J, 10 . . . 2 (and so on through the other three suits) it would be considered a very strange coincidence, yet any *particular* sequence is as likely, or rather as unlikely, as any other. Indeed, even the initial sequence of five, contemplated first and indicated explicitly (C.-9 ; S.-2, J ; H.-5 ; C.-A.), has a fair chance of being produced only on several hundred millions of trials ($52 \times 51 \times 50 \times 49 \times 48 = 311,875,200$). The meaning in calling a random sequence more probable than a well-ordered one is only that there is a host of sequences that we term random against only a few well-ordered ones. Hence, by shuffling, a random sequence is usually turned into a random sequence—and so is a well-ordered one.

The probability of the state of a material system is assessed by counting the number of ways its ultimate constituents could be regrouped without altering the observed state of the system. This number is always " more than astronomically " vast. To ponder the likelihood of a situation that has actually occurred may seem pointless. The point is, however, that thus we learn something about the future. There is, namely, the general and very understandable rule that the system, when left to itself, will (nearly) always go over into a more probable state. This accounts for one of the most fundamental laws of physics, the Second Law of Thermodynamics, which enounces that the entropy of an isolated system (nearly) always increases. Hence the prominent rôle played in physical science by statistics, in the sense of counting how many different arrangements of ultimate particles would realize the same observable situation.

In the last twenty-five years or so it has become clear that the counting, as previously performed, was inadequate, and had only by good luck, as it were, given correct results in many cases. The wrong way of counting is now termed " classical statistics ". Of the correct way there are two alternatives : termed Bose-Einstein, and Fermi-Dirac, statistics respectively. But it is not a matter of eliminating a gross error, in which case the amendment would be valuable, but far less interesting than it is. In fact classical statistics was blameless inasmuch as it was the only consistent procedure to adopt, considering what the ultimate particles were then deemed to be. Not the statistical considerations were at fault, but the idea one had formed about the objects to which they were to be applied. The idea of the " atoms ", taken over from Democritus (about 460 B.C.), had retained all through the centuries, until quite lately, a fundamentally wrong feature.

This feature and its bearing on statistics can be explained by three simple examples, illustrating the three kinds of statistics respectively. Supposing we know that in a tennis tournament two trophies (say the " singles open " and the " singles handicap ") have gone to the group of five Englishmen who competed, but we do not know to which of them : how many different possibilities are there ? Obviously there are twenty-five, since there are five different allocations for each of the two trophies, and every combination is possible and represents a different case (classical statistics). In the second example we let the lucky events that have descended upon our group of five men be two prizes in the same lottery of £100 each. Here the first way of counting would be at fault because it would mean counting the cases when the prizes go to two different winners twice over. Now there are five cases in which a single winner gets both ; the remaining twenty cases have to be halved ; hence the correct count is five plus ten, that is, fifteen possibilities (Bose-Einstein statistics). Thirdly, we take for the lucky events two knighthoods. This does away with the five cases in which one man got both and leaves us with only ten possibilities (Fermi-Dirac statistics).

To-day we know that some kinds of particle (photons and mesons) have to be dealt with statistically according to the second example (money), most of them (electrons, protons, neutrons) according to the third example (knighthoods), none according to the first (individually distinct trophies). This goes to show that the very nature of a particle excludes individual identification—a feature that hitherto, ever since the Greek atomists had conceived the idea of ultimate particles, had been taken for granted without giving it any thought. Not that particles were believed to carry identification marks in the manner of playing cards or trophies, but it was deemed compatible with their nature that they might. Our similes could not be adequate if this were so. Indeed, if a cheque is paid into a bank and the account is drawn on the next day, it is drawn on the balance, not just on the previous day's credit. With knighthoods there is not even a balance, only a " Yes " or " No ". Either one " belongs " to the K.B.E.s, and so represents (that is the meaning of the simile) a state occupied by an electron ; or one does not " belong " to them, thus representing a state which might be occupied by an electron but actually is not.

This last simile adequately renders W. Pauli's famous exclusion principle, which is closely allied to the Fermi-Dirac statistics, and consists precisely in excluding " double occupation " : there can never be two or more such particles (*e.g.* electrons) in the same state. The simile could be varied in many ways. Knighthood could be replaced by membership of almost any club, society, nationality, or creed. One could also vary the carriers, representing the states of the particle ; the particle must then be represented by a characteristic capable of two specifications only ; for instance the various states might be represented by the street lamps of a town, the particle by the characteristic " being lighted ". It is fairly obvious that entities which have to be symbolized in some such manner, if one wishes to understand their actual statistical behaviour, cannot aptly be pictured as tiny specks of matter, as they formerly were.

The new view amounts to a certain dematerialization of the concept of a particle, which is by no means easy to visualize. It is, however, supported by the observed fact that particles are occasionally " created " or " annihilated ", though not out of nothing or into nothing, but concomitantly to the annihilation or creation, respectively, of some other kind of particle, or to the loss or gain of energy in some other system. For instance, a light-quant (or " photon ") disappears when it is absorbed by an atom, which is thereby lifted into a higher energy level. Alternatively a strong photon may disappear and create a pair of electrons, a positive and a negative one.

The experimental evidence on which the new views are based is vast and multifarious, and derives its convincingness from an intricately interwoven network of inferences which can only be skimmed here. The exclusion principle gives the clue to the periodic table of the chemical elements and to the structure of their line spectra. It has also, in the hands of F. London and W. Heitler, elucidated the nature of that type of chemical bond which is prevalent in organic compounds and does not simply consist in the electrical attraction between oppositely charged atoms or ions. The properties of metals, to wit, their high conductivities for electricity and heat, their specific heats, their electronic radiation at high temperature (Richardson effect), and other phenomena seemed, under the aspect of classical statistics, to present glaring contradictions which were at once dissolved when A. Sommerfeld dealt with the electrons inside the metal according to Fermi-Dirac statistics.

The peculiar traits (called " degeneracy ") by which the behaviour of an assembly of particles differs from what classical statistics would let us expect are imperceptible at low particle density or at high temperature ; they become ever more prominent as either the density is increased or the temperature lowered, or both. Degeneracy prevails in the interior of some stars, in spite of the high temperature, because, as the centre of the star is approached, the weight of the outer layers compresses the material to extraordinarily high density, up to 100,000 times the density of water, or more than a ton per cube inch. In the laboratory, liquid helium at a few degrees above absolute zero exhibits unprecedented mechanical and thermal properties, which are no doubt mainly degeneracy. The helium atoms are controlled by Bose-Einstein statistics. So are the light-quanta (photons) that constitute the so-called black-body radiation.

However, in this famous problem, which led Max Planck in 1900 to the discovery of the quantum of action, classical statistics happens to be entirely correct if the radiation is pictured not as an assembly of photons, but as an electromagnetic wave phenomenon. And since the latter aspect had here an historical priority of long standing over the particle aspect, this case was at first less impressive. In all other cases the wave aspect is a novel conception of only about twenty-five years ago. It furnishes the simplest general formulation of the new statistics, which it reduces to classical statistics applied to other objects (not the particles). To explain this theory would exceed the scope and the limits of this report.

BOOK IV

SCIENCE AND EVERYDAY LIFE

SCIENCE AND EVERYDAY LIFE

SCIENCE has made immense progress in our lifetimes, and it will undoubtedly continue to do so. We sometimes see pathetic and quite fruitless appeals in the press for scientists to " take a holiday ". That is quite impossible. Whether we like it or not, science is going to keep galloping along, and we must face that fact either with approval if we think it a " good thing " or with courage if we don't. In this last book we shall return to our usual rallying point, and thus we shall see whether or not we shall have to call on our reserves of courage.

And first of all let us ask: *What has given science its power in our world ?* The answer is given in the following inspiring talk by Dr. Bronowski.

J. BRONOWSKI

A SENSE OF THE FUTURE

SEVENTY-five years ago, if you had walked on a summer evening into the country just beyond Bromley in Kent, you might have come on a remarkable sight. In the greenhouse of one of the larger and uglier houses of the neighbourhood a tall man in his sixties was stooping over potted plants. Beside him sat a younger man, just as absorbed, and the young man was playing the bassoon. This earnest pair were Charles Darwin and his son Frank ; and they were making a scientific experiment. Darwin wanted to know exactly what tells an insect-eating plant like the common sundew to close its leaves when a fly settles on it. So he was going through the possible causes methodically one by one. Noise was not a likely cause ; but it might just have worked ; and Darwin was not the man to rule out anything. He had tried sand and water and bits of hard-boiled egg, and now he was trying Frank's bassoon. Darwin never did get to the bottom of what makes the sundew close. But he almost did, and the next generation finished his work. He was well content with that. Darwin at sixty was a famous scientist who had changed our whole understanding of nature ; yet he remained content to do tidy experiments that would bear fruit somewhere, some time in the future.

This is the sense of the future I want to talk about, at first hand, as a scientist. I am distressed to see how many people to-day are afraid of the future and of science together. I believe that these fears are mistaken. They seem to me to misunderstand the methods of science ; and spring from a gloom about what it has done, which has simply forgotten the facts. We sit under the shadow of the nine o'clock news, nursing our sense of doom, and we think ourselves worse off than our forefathers a hundred and fifty years ago, who were at war with Napoleon for a generation. But a hundred and fifty years ago the working week was eighty hours for children. Cholera was more common in England than 'flu. The country could barely support ten million people, and not a million of them could read. You know how all this has been changed ; and don't let anyone tell you that nothing has been gained but comfort. Think of the gain in life and health alone : a population which has just topped fifty millions, the infant death rate cut by eighty or ninety per cent, and the span of life enlarged by at least twenty-five years. The sewer and the fertilizer have done that : and the linotype and the X-ray tube, and the statistician puzzling over inheritance. They have been real liberators. Every machine has been a liberator. They have freed us from drudgery and disease and ignorance and from the misery Hogarth painted that could forget itself only in the stupor of drink.

We owe that miracle to science ; and it is a miracle. But the scientists who have worked it have been neither gods nor witch-doctors. They have been men : men who had faith in the future ; and they have used no magic. What they have used is at bottom only Darwin's method ; because that method is science. Science is experiment ; science is trying things.

It is trying each possible alternative in turn, intelligently and systematically ; and throwing away what won't work, and accepting what will, no matter how it goes against our prejudices. And what works adds one more piece to the slow, laborious, but triumphant understanding of our world.

This is not a secret or a mysterious progress. If it sometimes seems so, that is just because the day-to-day work of science is so unspectacular. You hear nothing from the research worker for years, and then, suddenly, there is the result in the headlines : penicillin or the jet engine or nuclear fission. No one tells the layman about the years of experiment and failure. How is he to know what has not been done, or to guess the labour of what has ? What is he to think but to marvel at the skill of science, and to fear its power ?

I believe that both these feelings do equal harm : the feeling of marvel as much as the fear. Because they have this in common that they both want to persuade the layman that there is nothing he can do for himself. Science is the new magic, they whisper ; it is out of your hands ; for good or ill, your salvation or your doom is the business of others.

That is why I have attacked the magic before the fear : because the marvel lies below the fear. In the minds of most people to-day, the fear is plainly uppermost. They are afraid of the future ; and if you ask them why, they conveniently blame the atomic bomb. But the atomic bomb is only the scapegoat for our fears. We are not afraid of the future because of a bomb. We are afraid of bombs because we have no faith in the future. We no longer have faith in our ability, as individuals or as nations, to control our own future. That loss of confidence has not sprung overnight from the invention of a weapon. The atomic bomb has merely brought home to us, harshly, as a matter of life and death, what has long been growing : our failure to face, our refusal to face, as individuals and as nations, the place of science in our world.

There is the taproot of our fears. In our hearts, of course, we know that the future belongs to science ; we do not deceive ourselves about that. But we do not want to have to think like scientists. We want to cling to the doctrines and prejudices which we imagine, quite wrongly, made the world snug fifty years ago. We do not care about the future ; we just want that world to last our time. Because we do not feel equal to the new ideas ; we have been told that science is mysterious and difficult. And so we let the exciting new knowledge slip from us, a little further every day, and our confidence with it ; and then, face to face with the sense of our helplessness, we pretend that it is all a conspiracy among nuclear physicists.

It is in our power to change that in our own generation. As nations, we can apply to affairs of state the realism of science : holding to what works and discarding what does not. As individuals, we can grasp the common-sense ideas of science. And there is the most important lesson we must learn : it is the ideas of science that are remaking the world, not its mechanical achievements. When we have learnt that, we will see the achievements too in their proper place. The atomic bomb is not a great achievement of science. But scientists made a great discovery : the fundamental discovery that we can tap atomic energy. That is an achievement not of bickering nations but of man. And we have the whole history of science to tell us that every fundamental discovery has in the end brought men more good than harm. I said " has, in the end " almost by habit : has, if we are willing to look forward. Every scientist looks forward ; what else is research but to begin what others will finish and enjoy ? And what other incentive can satisfy any of us but that sense of the future ? Disaster threatens us only if we perpetuate the division between science and our own everyday living and thinking. Let no one tell you again that science is only for specialists ; it is not. It is no different from history or good talk or reading a novel ; some people do it better and some worse ; some make a life's work of it ; but it is within the reach of everybody. Science is as human as Darwin and his bassoon, and no harder to understand. Its values are the human values : honesty, tolerance, independence, common sense, and singleness of mind. Its achievements are among the great achievements of man : the Greeks ranked Pythagoras with Homer. And it has made its way not secretly and not by authority but by sticking to the plain facts and only the facts—never mind who discovered them or who challenged them. Science listened equally to Newton and his friend Christopher Wren, to Darwin and his critic Samuel Butler : and listens to-day to every bright lad with an idea as patiently as to the professors. If you want to know what happens to science when it allows itself to be dominated by authority, political or scientific, let me take you to a field of which I have some special knowledge : German research during the war. We went into the war very much afraid of German science : it had once had a great reputation. Yet the Germans all through the war never took a fundamental step, whether in U-boat research, in radar bombing, or in nuclear physics. Why were they, the professional war-makers, outclassed by us? One example will tell you. About the time that we had our first atomic pile working, Himmler's director of war research was sending an investigator to Denmark to discover—believe it or not—how the Vikings knitted. By one of those exquisite strokes of irony which dogged the Nazis, the name of this investigator—believe it or not—was Miss Piffl.

To listen to everyone ; to silence no one ; to honour and promote those who are right : these have given science its power in our world, and its humanity. Don't be deceived by those who say that science is narrow ; a narrow, bigoted power is as brittle as Himmler's. Have you been told that science is dogmatic ? There is not a field of science which has not been made over from top to bottom in the last fifty years. Science has filled our world because it has been tolerant and flexible and endlessly open to new ideas. In the best sense of that difficult word, science is a democratic method. That has been its strength : that and its confidence that nothing can be more important than what is true.

Does that seem to you, after all, a very ordinary tradition ?

Of course it is. It is the tradition for which Europe has hankered ever since the Renaissance : free inquiry and personal action. It is the climate of the arts as much as of science. England led the world in both, because from Elizabethan times she made that tradition of independence her actual way of living. That is why the Authorized Version of the Bible, and the first table of logarithms and Shakespeare's Folio, all came out in England within twelve years. It is our inheritance of freedom, which has liberated the mind with the body. The sense of the future and that tradition are one, if we are willing to unite them. The ideas of science are not special ideas ; we can all get at the heart of them—that is, all of us who are willing to find Darwin's sundew more stirring than the Vikings. What we need is to stop shutting our minds to these ideas ; to stop being afraid of them. We stand on the threshold of a great age of science ; we are already over the threshold ; it is for us to make that future our own.

" Science is a democratic method." " That " says Dr. Bronowski " has been its strength ; that and its confidence that nothing can be more important than what is true." Certainly science does not seem to flourish in the totalitarian state. Science suffered in Germany under Hitler, and it is suffering in Russia under the Communist régime. But Germany suffered as well as science, and Russia will no doubt suffer in much the same way. So there must be something more fundamental than method, whether it be democratic or totalitarian, and we have this in Dr. Bronowski's second factor—the devotion of science to what is true.

Truth is all-important to science, and we have here a vital link with literature, or at least with the highest forms of literature and the arts. *Why is science worthy of study by poet and novelist ?* You will remember that Ball poked gentle fun at Wolfe's poem on the Burial of Sir John Moore because the poet had not consulted a calendar before introducing the moon into his picture. That is not a serious criticism ; the poet has the artist's licence to introduce or to suppress little details so as to give greater unity and life to his picture. But the case is different when the writer gives grossly false impressions which are not true to Nature. The pursuit of truth should be the occupation of both man of science and man of letters. Lord John Russell realized this and brought it home to his audience in the following address, delivered a century ago.

LORD JOHN RUSSELL

SCIENCE AND LITERATURE AS MODES OF PROGRESS

BEFORE the Reformation, and immediately afterwards, great sums of money and land were given for the purposes of endowing academies, colleges, and schools for education. Our ancestors thought, and I believe wisely thought, that the best plan they could adopt was to teach, or to provide means for teaching, the science and literature which had been derived from ancient nations; for in those days that science and that literature contained all that was known, and was really worthy of study, the most profound works, upon subjects of geometry and science, and the best models of literary writing. I am far from thinking that our ancestors committed an error, either, when they directed the education of youth almost exclusively to these objects, or when they decided that a great length of time should be given to that knowledge; but we have to consider that in the present day we stand in a totally different position. Not that we ought to forget what great advantages we have derived from the science and the literature of ancient nations; because upon the geometry delivered to us from the ancients has been founded all that increase of knowledge which ended in the discoveries of Newton; from the writings of the poets of antiquity the great poets of modern times have derived the best models they could imitate; from the jurisprudence of the Romans were derived the laws by which most of the nations of the Continent have been ruled.

But, while this tribute must be paid, it is a paramount object of attention that we, in the course of the three centuries and a half that have elapsed from what is called " the revival of letters ", have added to the stores that we have received immense stores of our own—that by the side of that rich mine we have opened other mines, which, if not of richer ore, are more easily worked and more abundant in their produce. It was Bacon who first pointed out that the mode of the pursuit of science for modern nations ought to be different from that mode for the discovery of truth which has been pointed out by some of the great philosophers. It has been much questioned whether Bacon was in fact the guide by whom other discoverers have been enabled to pursue the track of knowledge and of invention, and upon that point I think it is certainly clear that it was not Bacon who enabled Galileo and Torricelli, Pascal, Tycho Brahe, Copernicus, and Kepler to make the great discoveries which have immortalized their names.

But what is true is that Bacon at a very early period laid down the rules by which all modern men of science have guided themselves. He pointed out the road they have followed, and laid down more clearly, more broadly, more ably than anyone else, the great method by which modern discovery should be pursued. You will find, I think, if you pursue this subject—if those who belong to mechanics' institutes will study the two works of Bacon, the one called the *New Organ* and the other *The Instauration of the Sciences*—you will find that the latest discoveries, the latest inventions, have been made according to that mode which he pointed out. A work was published but a year ago by Mr. Fairbairn, giving an account of the experiments which he adopted under the direction of Mr. Stephenson, and by which that gentleman was enabled to construct the tubular bridges at Conway and over the Menai Straits. You will find that all those experiments were according to the rules which Bacon has laid down.

Take another on geology, and a most interesting work it is, called the *Old Red Sandstone*, by Mr. Hugh Miller, and you will find in that interesting work, which is as remarkable for the beauty of its style as for the importance of its matter, that Mr. Hugh Miller, being at first a mason working in a stone quarry, pursued, in his method of investigation, the same rules which Bacon more than three centuries ago laid down, and which have thus become the foundation of the law, as it were, of modern science.

And now, ladies and gentlemen, having said this much with regard to the original method, let me venture to say that, interested as no doubt the members of the Mechanics' Institute may be in the various sciences which of late have made so great a progress—that, interesting to you as are those discoveries which have given us the power of rapid locomotion and the electric telegraph—wonderful and extraordinary as all those discoveries are, and the study of the means and methods by which they were made, I would earnestly press upon you that there is one science which, though its practical use is rather upon the sea than upon the land, is yet worthy of the deepest study, on account of the magnificent results which it unfolds. The science to which I allude is the science of astronomy.

Whether those who, having begun the mathematical studies with the simplest problems of geometry, wish to pursue them to the end, and follow the works of Newton himself—and no more interesting works can be studied by a mathematician—with the view of seeing how it was that he discovered that great law of gravitation by which his name will be for ever known, or whether, contenting yourselves with the popular accounts of astronomy in many of the works of the day, written by Sir John Herschel and other eminent men—whether you pursue one branch or the other—you cannot fail to be struck with the dread magnificence of heaven which is unfolded to you in astronomical speculations. That course of discovery, be it remarked, is still open—it is still pursued ; and it is but lately that it has been found that those parts of the heaven which seem to be mere collections of luminous clouds, and not to contain anything like form of world or form of suns, are in fact full of stars, small in appearance to us, but really of very great magnitude, though at an immensely remote distance ; so that, as it were, a new heaven is opened to us, and it appears that to Him to whom " a thousand years are but a day " a thousand worlds are but a speck.

I certainly shall not attempt to detain you and to occupy your time by speaking of any of those other sciences which all have their delight and their utility. . . . I will now turn for a short time to the subject of literature. That subject again is so vast that if I were to attempt to go over any one of its numerous fields I should not find the time sufficient to enable me to do so ; but there is one leading remark which I will venture to make, and which, I think, it is worth while for any person who studies literature to keep in view. There are various kinds of productions of literature of very different forms and of very different tastes—some grave and some gay, some of extreme fancy, some rigorously logical, but all, as I think, demanding this as their quality—that truth shall prevail in them.

A French author has said that nothing is beautiful but truth ; that truth alone is lovely, but that truth ought to prevail even in fable. I believe that remark is perfectly correct ; and I believe that you cannot use a better test, even of works of imagination, than to ask whether they be true to Nature. Now, perhaps, I can better explain what I mean in this respect by giving you one or two instances than I should be able to do by precept and explanation. A poet of very great celebrity in the last century, who certainly was a poet distinguished for much fancy and great power of pathos, but who had not the merit of being always as true as he is pointed in the poetry he has written—I mean Young—has said, at the commencement, I think, of one of his " Nights " :—

> " Sleep, like the world, his ready visit pays
>   Where Fortune smiles ; the wretched he forsakes,
>   And lights on lids unsullied with a tear."

Now, if you will study that sentence, you will see there are
two things which the poet has confounded together. He has
confounded together those who are fortunate in the possession
of health, and those who are fortunate in worldly advantages.
Now, it frequently happens that the man who is the worse off
in his worldly circumstances—to whom the world will pay no
homage—on whom it would not be said that Fortune smiled,
enjoys sweeter and more regular sleep than those who are in
the possession of the highest advantages of rank and wealth.
You will all remember, no doubt, that in a passage I need not
quote, another poet—one always true to Nature—Shakespeare,
has described the shipboy amidst the storm, notwithstanding
all the perils of his position on the mast, as enjoying a quiet
sleep, while he describes the king as unable to rest. That is the
poet true to Nature ; and you will thus, by following observations of this kind, by applying that test to poetry as well as to
history and to reasoning, obtain a correct judgment as to
whether what you are reading is really worth your attention
and worth your admiration, or whether it is faulty and is not
so deserving.

I may give another instance, and I could hardly venture to
do so if my friend Lord Carlisle were here, because the want of
truth I am going to point out is in the writings of Pope. There
is a very beautiful ode of Horace, in which, exalting the merits
of poetry, he says that many brave men lived before Agamem-
non ; that there were many great sieges before the siege of
Troy ; that before Achilles and Hector existed, there were
brave men and great battles ; but that, as they had no poet,
they died, and that it required the genius of poetry to give
immortal existence to the bravery of armies and of chiefs.
Pope has copied this ode of Horace, and in some respects has
well copied, and imitated it in some lines which certainly are
worthy of admiration, beginning :—

> " Lest you should think that verse shall die,
>   Which sounds the silver Thames along."

But in the instances which he gives he mentions Newton, and says that not only brave men had lived and fought, but that other Newtons " systems fram'd ". Now, here he has not kept to the merit and truth of his original, for, though it may be quite true that there were distinguished armies and wonderful sieges, and that their memory has passed into oblivion, it is not at all probable that any man like Newton followed by mathematical roads the line of discovery, and that those great truths which he discovered should have perished and fallen into oblivion. I give you these two instances of want of truth even in celebrated poets, and I think it is a matter you will do well to keep in view, because there is a remarkable difference between the history of science and the history of literature.

In the history of science the progress of discovery is gradual. Those who make these discoveries sometimes commit great errors. They fall into many absurd mistakes, of which I could give you numerous instances ; but these blunders and these errors disappear—the discoveries alone remain ; other men afterwards make these discoveries the elements and groundwork of new investigations, and thus the progress of science is continual ; but truth remains, the methods of investigations even are shortened, and the progress continually goes on.

But it is not so with regard to literature. It has, indeed, happened often in the history of the world, among nations that have excelled in literature, after great works had been produced which brought down the admiration of all who could read them, that others, attempting to go further—attempting to do something still better—have produced works written in the most affected and unnatural style, and, instead of promoting literature, have corrupted the taste of the nation in which they lived. Now, this is a thing against which I think we should always be upon our guard, and, having those great models of literature which we possess before us—having Shakespeare, and Milton, and Pope, and a long line of illustrious poets and authors—we should always study to see that the literature of the day is, if not on a par with, at least as pure in point of taste as that which has gone before it, and to take care that we do not, instead of advancing in letters, fall back and decay in the productions of the time.

I will now mention to you another instance. It is apparently but a trifling one, but still it is one in which I think Nature and truth are so well observed that it may be worth your while to listen to it. One of our writers, who the most blended amusement with instruction, and ease of style with solidity of matter, as you all know, was Addison. He describes a ride he had along with a country squire, whom he fell in with in travelling from London to a distant town. They came to an inn, and Addison says that they ordered a bowl of punch for their entertainment. The country squire began, as was, perhaps, a mode with country squires, which may have continued even to the present day, to deprecate trade, and to say that foreign trade was the ruin of the country, and that it was too bad that the foreigner should have so much advantage of our English money. " Upon which " says Addison " I just called his attention to the punch that we were going to drink, and I said : ' If it were not for our foreign trade, where would be the rum, and the lemons, and the sugar, which we are about to consume ? ' " The squire was considerably embarrassed by this remark, but the landlord, who was standing by, came to his assistance, and said : " There is no better drink than a cup of English water, if it has but plenty of malt in it ". Now, although that appears a slight and trifling story, and told in a very common way, yet it is perfectly true to Nature, and it conveys in a lively manner a rebuke to the ignorance and prejudice of the person with whom Addison represents himself to be conversing.

Having made these observations, you will, perhaps, permit me, ladies and gentlemen, to say that the cause of my venturing to come here is, that I might both see the progress that you are making in instructions of all kinds, and also that I might express my hopes and my wishes for your welfare in the time that is to come. It has been my fortune, since the active part of my life began, to live in times of peace and to see great discoveries and great improvements. I think you will feel that we who have had the direction of affairs during that time—I speak not now of any differences of political parties or of religious sects, but taking us altogether, all political parties, and men of all religious denominations—I think that we have not done ill for the country during that period in which we have borne an active share in its affairs. If you look back to 1815, when a bloody and costly struggle terminated, I think you will see that since that period, whether by judgment of Parliament —whether by the action of great bodies and societies—or whether by the skill and invention of individuals, the condition of the people of this land has very much improved. While the means of sustenance have become cheaper—while the public burdens have become less—while the means of education have been improved—there has been, with these circumstances, and partly owing to these circumstances, a general progress in society.

I think that we who have belonged to that time—and, as I tell you again, I wish to make no political allusion, or to claim for one party over another any advantage—but I say generally that we who have lived in this time have, upon the whole, not ill performed our duty. It will be for you, when we retire from the more active business of this scene, to endeavour to carry on to a still greater knowledge, to still more comfort, to still greater well-being, the country in which you live. There is a great charge imposed upon you, and I trust you will properly perform it. Let no insane passion carry you without reason into contests with foreign countries. Let no unworthy prejudices induce you to withhold from any part of your countrymen that which is their due. Let no previous convictions prevent you from examining every subject with impartial eyes, and from placing before you the light of truth, which ought to guide you in your investigations. With these convictions I am persuaded you will abide by the institutions which you have, by the faith which you hold, and that you will adorn the country to which you belong.

It was in the nineteenth century that science first began to play a major part in everyday life. *What changes did science produce in nineteenth-century Britain?* Let us consult E. F. Armstrong's inaugural address to the Royal Society of Arts. He gives us a story of immense progress but also shows us how, after the initial success of Perkins in the discovery of aniline dyes, Great Britain allowed the German industrialists to take the lead in organic chemistry—a tragedy of wasted opportunity.

E. F. ARMSTRONG

SCIENCE IN NINETEENTH-CENTURY BRITAIN

WE are meeting within a few days of the centenary of the founding of an enterprise which owed much to the Prince Consort, namely, the beginning of the Royal College of Chemistry which to-day survives as part of the Imperial College of Science. The Royal College was unique in that it was established so that men should study chemistry at the laboratory bench, hoping to make discoveries to enlarge the borders of their science. They were going to make their careers and perhaps their livelihood out of chemistry, an experimental science which, in due course, would perhaps make contributions of benefit to mankind. The story of its formation has fortunately been documented by the Chemical Society in a series of lectures given to commemorate the life and work of Hofmann, first professor at the college.

Before the foundation of the Chemical Society in 1841 the profession, as such, can scarcely be said to have existed, although there had been a few practising chemists as well as dilettante chemical philosophers who were devoted to the science. When reviewing the history of the Chemical Society fifty years later Dr. W. J. Russell said that " the number of real students in 1841 was very small. They were looked upon by their friends as being eccentric young men who would probably never do any good for themselves."

A word or two is desirable as to the background against which the introduction of Science in Britain was taking place. One hundred years ago there were only fifteen million people in England. There were two million people in the City of London and there were less than four million people in all the towns of England, including London ; that is to say, England, so far as towns were concerned, was London. It was a time of change. The steam engine was beginning to be generally applied, railways were spreading the length and breadth of the land. Many of the railways were formed to carry minerals, the mining of coal was becoming of great importance as was its cheap and rapid transport to the user. It was an age of wheels, turning wheels in the cotton and woollen factories, wheels at the pit head to raise the coal from below, wheels to pump water from the mines in Cornwall. The market towns were building factories and the agricultural labourers were coming to town to work in them.

The battle of the railways with the roads was not entirely won till 1844, though the Liverpool-Manchester line had been opened with much pomp in 1830 and the Great Western Railway Bill enacted in 1835. The Queen made her first journey in 1842 ; Victoria Station, the London terminus of the new South-Eastern Railway, was so named in 1846. The railway boom, after a start in the middle 'thirties, attained its height in 1845, when new enterprises, many of which never came to fruition, were floated all over the country. By 1844 stage coaches were finally driven off at least the southern roads. In this year also, Gladstone introduced the cheap trains Act, for, since the Government had not contributed a penny to railway enterprise, it could hardly claim the control exercised over them in more enlightened countries. His Act gave the third-class passenger legal status and it was soon found by the companies that the cheap traffic they had disdained was the most profitable of all the passenger business.

The greatest of all reforms which was going on at this time was Rowland Hill's proposal for the conveyance and delivery of letters throughout the length and breadth of England at a uniform rate of one penny. The change gave an enormous impetus to communication all over Great Britain, both for commercial and all other purposes; overnight, what had been difficult and costly became easy and cheap. The Post Office in those days took very great pride in seeing that letters were delivered with the utmost possible speed. There were six collections and six deliveries every day in London, and it was usual for a person who posted a letter in the morning to receive an answer to it before going to bed that night. How right our ancestors were to give the first Albert Medal to Rowland Hill. Shortly afterwards the adhesive postage stamp was introduced on May 6, 1840; an unexpected result was that the excellent engraving of the Queen's head on the stamp brought the Sovereign into closer relation with her subjects all over the globe. It made the Queen real to people who had not been particularly interested in royalty, and loyalty to the Crown, which had been very largely absent during the days of the Hanoverian kings, began to be established. One of the results of that was the gradual return of power to the Sovereign as a constitutional monarch, power being taken out of the hands of the large families who were always fighting one another under the names of Whigs and Tories, so that our politics in those days were a struggle for power between families and not the kind of politics that we developed later. It would be wrong to omit mention of how bad the conditions of the poor were at this time. They are set out austerely in the second chapter of Sir George Arthur's book *Concerning Queen Victoria and her Son*, and in regard to children more vividly in Charles Dickens' *Oliver Twist*. Looking back, it is strange to think the country allowed children under five to be employed in mills and factories. I may recall the Society's own work in the introduction of mechanical means of sweeping chimneys and the prevention of child labour. Reformers had a hard time in those bad days but, when possible, they had Prince Albert on their side.

Hitherto the County Squire had reigned supreme, his position was now threatened by the industrialist who was making money fast ; new forces were coming into politics.

There was a stir, too, in educational matters and an attempt to spread knowledge more widely. The tempo of transport both for goods and for men had changed, the towns in the Midlands and North were becoming less remote from London. Hence, when Baron von Liebig, the outstanding German Professor of Chemistry and research worker, came to England in 1842, it was found both desirable and possible to send him on a tour of the industrial Midlands and North. He was conducted by Lyon Playfair, who was at pains to see that Liebig met the people who mattered and that he impressed on them the importance of the new science of experimental chemistry. It is on record that the tour was a triumph so that the merchants and industrialists were duly impressed, though apparently Liebig discovered that "England is not the land of Science ". . . .

As a consequence of this external stimulus movements were set on foot to encourage chemistry and its study in our island. Confining our attention to London, the first effort of the enthusiasts there was to attempt to found a Davy College of Practical Chemistry within the Royal Institution. For various reasons this did not succeed and after further meetings it was decided to found a new institution, the Royal College of Chemistry, on July 29, 1845.

Going back for a moment in chemical history, two self-made men, Davy and Faraday, did much to make chemistry and physical science popular by their charm and eloquence as lecturers. But later only Graham thought of opening his laboratory for the training of students in methods of research. Playfair was one of these students both in Glasgow and at University College, London, so that he was well prepared to play a part when the new movement came after 1842. The Pharmaceutical Society claim that the first laboratory open to students was theirs with Fownes as teacher. University and King's College gave increased attention to practical teaching and the Government School of Mines with an excellent laboratory was opened in Jermyn Street with Playfair to teach chemistry.

In Liebig's laboratory in Giessen about the year 1840, a school of chemists had collected from all nations, especially from England. Frankland regarded this as the first laboratory of the kind, and there is little doubt that no earlier laboratory had been better equipped or controlled by a better master.

But our interest centres on the Royal College of Chemistry since two of those most actively concerned in its foundation were the Queen's Physician, Sir James Clark, and the Prince Consort. The necessary subscriptions were obtained, temporary laboratories were found in George Street, Hanover Square, and A. W. Hofmann was secured as Director on the recommendation of Liebig. There were difficulties to be overcome owing to Hofmann's desire to be able to return to Bonn if the London venture failed, but the Prince Consort, then staying with Queen Victoria at Bruhl, intervened actively and a satisfactory arrangement was made. And so Hofmann started in London in October 1845.

The foundation stone of the new college was laid by the Prince Consort on June 16, 1846, and work in the new laboratory with accommodation for fifty students was actually commenced in October of that year.

The new venture was almost immediately successful. Led and inspired by Hofmann, the students who came during the first few years nearly all became famous leaders in the new chemical industries or in the application of chemistry to the arts. At rare moments, here and there in human history, in men and times of abounding vitality, the golden hour of progress is at hand. In the Royal College laboratories existence must have seemed beyond a doubt desirable. In Elizabethan days, the golden age of adventure, England's greatness was built by audacious and perhaps not over-scrupulous men with their trading, privateering, and warring.

" Westward Ho for Trinidad and Eastward Ho for Spain ! "

Now test tube, beaker, and flask replaced culverin and pike ; the audacity was there but the war was against nature's secrets and not against Spain.

The outstanding contribution of the Hofmann school was the foundation of the organic or aniline dye industry. At this period all colouring matters were of vegetable or mineral origin. Aniline had been discovered for the first time in 1841 by Fritzsche as a base produced by heating indigo, and it was by this laborious and wasteful process that Hofmann made it. A little later he separated it from coal-tar naphtha by successive nitration and reduction, and set his brilliant pupil, Mansfield, to undertake the difficult task of investigating the hydrocarbons present in this material. A most important step was made, for Mansfield obtained both pure benzene and toluene. Benzene had been first discovered by Faraday in 1825 in the condensate from the oil gas produced by heating fish oils. Aniline thus became more available to the chemist and formed the starting material from which W. H. Perkin in 1856 made the first synthetic dye—aniline purple or mauve. He was only seventeen at the time and working in his private laboratory at home, his aim being to make quinine. Instead, what appeared to be a dyestuff was obtained on which Perkin obtained a favourable report from Pullars of Perth. He patented the process, left college and, backed by his father and elder brother, set out to manufacture aniline purple. The French patent had been badly drafted and proved invalid, so French manufacturers were able to manufacture the colour in competition—they gave it the name of mauve.

The discovery attracted so much attention that the chase for colours was on. Nicholson made magenta and the rosaniline colours, whilst Perkin's Greenford Green factory extended their range ; in 1869 Perkin achieved his second great success in the synthesis of alizarin, hitherto derived entirely from the madder root.

Many of Hofmann's students became famous in the industry —not forgetting Peter Griess, the discoverer of azo dyes, who settled down in a brewery—Allsopps—at Burton-on-Trent. Other Germans among his assistants, for example, Caro, Martius, O. N. Witt, went into the English industry but later returned to Germany to become founders of the German dye industry. Indeed, within the next ten years the German dye works forged ahead of their English competitors and maintained an ever-increasing lead over them. If I may be personal for a few moments, I should like to tell you that my father was Hofmann's last student. Hofmann was called back to Berlin in 1865 ; he left rather hurriedly in the end, and my father was under him for only a week or two, but I have here my father's admission card for the Session 1865-6, " to hear lectures on inorganic chemistry ". This card is interesting in relation to the centenary of the Imperial College of Science, which we celebrated a week or so ago, as it takes us back eighty out of the one hundred years. I have also here my father's card of admission for the next year—he had progressed from inorganic to organic chemistry during the course of a year—and I have half a dozen other cards for other subjects. The College was then called " The Royal School of Mines and Science applied to the Arts ".

There is no doubt that the dye industry languished here because, as C. J. T. Cronshaw has written, the pioneer spirit and the creative instinct which brought it into being abandoned it too early, little knowing that what they had accomplished was the merest scratch on the surface. It must be remembered, as M. O. Forster observes, that at this time British industry was blazing, and when viewed in this floodlight the stellar magnitude of Perkin's enterprise must have appeared negligible to British manufacturers : to the Rhine it was a crumb from the rich man's table.

At least no one realized that with dyes England lost to

Germany the organic chemical industry, ultimately to become one of the most lucrative and important of any. The synthetic organic chemical industry is still perhaps that industry with the greatest potentiality for the future and one which we must not dare to keep out of unless we would become a third-class nation. We are far stronger in Chemical Industry than in 1914, but the danger is still there largely because it is so hard to make the lawyers, accountants, and politicians who rule our destiny realize and understand either the nature, magnitude, or urgency of the problems of plastics, drugs, insecticides, and other organics !

Such people do not feel disposed to admit scientists to their kingdom. Bernal reminds us that men are slow to realize the importance of understanding what they are doing and having a theory behind their industrial activity. The idea of a perpetually changing industry digging out its foundations all the time is new and not one the older industries will readily accept. Cotton and coal are outstanding examples of the persistence of the attitude of mind of rule of thumb and hard business sense which, though it created the industrial revolution, is now the biggest brake on it. In Walter Page's phrase, " The winds of the world have not ventilated our brains ".

Ernest Bevin has asked for a scientific study of the problem how best to bring the benefits of scientific discovery into the lives of the people so that they can be enjoyed by the masses as quickly as possible and at a price within their reach. He has further said, " I think it obvious we could and should bear the cost of the development of scientific discoveries ". The people should, as it were, have a vested interest in scientific discovery. They would then be receptive to new ideas and welcome scientific progress.

But England was not yet converted to science in spite of the early enthusiasm of the Royal College of Chemistry. The names of Hofmann's later students are not so well known, which may indicate a falling off in the attractions of chemical research as a career. The Royal College, when it moved to

South Kensington, became more and more an institution for the training of teachers.

During most of the last hundred years the classical tradition has prevailed in our schools. The high road to distinction has lain through the humanistic schools of the older Universities, with the result that the control of industry and still more of the State policy has been largely in the hands of men with little understanding of how science goes to work and what it can accomplish. The scientist, if employed or used at all, has been relegated to a subordinate position and not permitted to make any real contribution to social and industrial affairs.

A few reformers, notably my father and the Educational Section of the British Association, fought for many years on behalf of science teaching in the schools, but with few exceptions it has remained as a belated and unwelcome part of the educational system and the new Education Bill has, I fear, not improved matters.

Contrast this attitude with Germany, where, for over seventy years, industrial leaders realized that it was good business to spend money on research. In fact, the 1914 war began at the very zenith of Germany's deservedly high scientific reputation. In almost every industry where science had been applied they led the world. When war came every German chemical factory was transformed into an arsenal. Our scientists rendered great service during the 1914-18 war but these were soon forgotten by the State. It is true there was a great increase in the numbers of scientifically trained men coming out of the universities during the period 1920-45, but the old traditions still prevailed, all the key posts going to men totally ignorant of and unable to comprehend science. Only the newer, more scientific industries gave opportunities and we were forced to watch once again the progress in Germany, and now in America also, whilst the State gave little encouragement to science. True, the Department of Scientific and Industrial Research did what was possible, and it is only during the war that it has been appreciated how much this was.

Once again, when at war, the country needed science more urgently than ever before. Led by a Prime Minister ably advised by Lord Cherwell, the nation has made full, though sometimes grudging and belated, use of its scientists and largely through them victory has been achieved, for this has been essentially a war of brains and mechanical weapons. The war is over, quickly in the end, through the use of the atomic bomb, a weapon of terrible import, but the discovery of British scientists. Science is officially patronized during war-time as the source from which convenient or necessary inventions may be expected, but will it remain in peace-time a mere minister to material ease and facilitated communications ?

Are we to be cast aside again whilst the lawyer and the accountant play havoc with our industries through sheer ignorance of what industry is ; or is scientific knowledge, scientific method, to be allowed to play a part in technical as well as in administrative matters ?

Quoting Bernal once more :

" The war has shown to millions of people in the services and in the war factories how important science can be. The very demands which are now being made for adequate food, housing, and health are known to be based on scientific studies. The new standards can be achieved only by the application of science. Beyond that, we are beginning to see that not only these problems but many political and social problems also depend on science for their solution. With this realization we may hope will come an increasing emphasis on the balanced development of science, on its increasing use in satisfying social needs, and on the spread into ordinary life of a scientific way of thinking and acting."

If history has any meaning, we are on a voyage hardly yet begun. We do well when we look back, we know not why, with instinctive fascination upon all the historical movements, refusing to sacrifice any one of them—neither the accumulated

wisdom nor the capitalized experience of the generations that
have gone before us.

The heroic heart supports the resolution : " To strive, to
seek, to find, and not to yield ". Mankind is committed to a
long journey. Knowledge and wisdom are of slow growth, as
history is witness, and the universe offers an extensive field
of inquiry.

If progress was rapid in the nineteenth century it has been even
more swift in the twentieth, as Prof. Andrade will show in the following
talk. *How has science made man's labour easier and his leisure more
abundant and more satisfying ?*

E. N. DA C. ANDRADE

SCIENCE THE SERVANT OF HUMANITY

IN dealing with some aspects of the very large subject of
science in the modern world I am going to try to be quite
simple. I shall not go very deeply into anything, but just
point out a few of the ways in which science is at work all
around, and how, whether you will or no, it touches your lives
at every point. I shall have to pick out just a few examples
that seem typical to me, and leave out hundreds of other
examples that would be, perhaps, just as good. I ought also
to say that I shall generally be speaking of the sciences of
physics and chemistry, in their various branches, and that I
shall say nothing of the medical sciences which play such a
large part in keeping us alive. This is not because I do not
realize that they provide splendid examples of science in the
modern world, but because we cannot discuss everything, and
also because I am not a medical man.

We may perhaps think of the work of science as being

carried out in two main ways. Firstly, there are the great men who, in the laboratory and study, make those wonderful discoveries that decide the whole course of a branch of science, or who, like Newton, map out the whole path of discovery for generations to come, and show them how to adventure successfully into the deserts of ignorance. There are men like Faraday, who discovered the connexion between electricity and magnetism on which all our modern electrical industry is built : Carnot and Joule, who gave us the laws which govern our heat engines, one working on paper and the other in the laboratory : Dalton, Lavoisier, and Berzelius, who organized the science of chemistry—but I shall not weary you with a list of names. Coming down to more modern times, we have men like Sir J. J. Thomson, from whose laboratory came all those early researches on the electron which have had such an influence upon the developments of the X-ray tube, the electric lamp, and the wireless valve. Men of this type work for pure love of knowledge, to find out how Nature works, and they are surrounded by groups of younger men working from more or less disinterested motives and with greater or lesser degrees of success. Of the discoveries of these giants you often read and hear.

Besides these, however, science has an army of men at work turning their discoveries to practical account. Among these stand out the great inventors, who point out that something useful from a practical point of view (or, as the more material-minded would say, something with money in it) can be built by applying the teachings of pure science. But this is only a first step. When a device has been invented, it has to be manufactured, and, generally, manufactured cheaply. Before this can be done there are in general a host of small scientific problems of great interest to be tackled—suitable materials to be found, processes to be worked out, reasons to be found for unaccountable failures in manufacture or use of the article in question. Here again science is at work, not only discovering how things can be done, but how they can be done systematically ; how they can be done safely ; and, above all, how they can be done cheaply.

The economic problem is always before us, and to-day one of the chief tasks of applied science is not to invent new industries, but to investigate the old ones and to see how and why the processes employed in them really work, and then to see how they can be simplified, controlled, and cheapened. So we find that every really successful industry has large research laboratories attached to it, in which men of science are at work on problems connected with the industry. Some of the more enlightened of these laboratories employ very great scientists, pay them exceedingly well, and let them do more or less as they like. This costs them immense sums. The Western Electric Company of America is supposed to spend about £2 million a year on these research laboratories. Why ? Because it pays. Someone asked a member of the research staff if there was any difficulty in securing from the company the funds required for research. " Oh, no " he said " it mostly comes in the form of royalties paid by foreign countries." Those foreign countries include, of course, England.

Besides various research laboratories connected with industry, like that of the General Electric Company at Wembley, we have in this country, you know, a Department of Scientific and Industrial Research—at least you ought to know, for you as taxpayers pay about £5 million a year for it. This department has working in close connexion with it some twenty research associations, supported by groups of manufacturers, such as the Cotton Research Association, the Woollen and Worsted Research Association, and so on. I am going to try to show you the kind of work which they are doing. It may be that I shall only very occasionally mention a research association by name, for lack of time, but when at various times I refer to problems of to-day, you will see the need and use of these associations.

SCIENCE IN THE HOME

Now let us turn for our first example to science in the home. When I first began to think about this talk, and was wondering where I should begin, I happened to be at breakfast, eating a boiled egg, and it occurred to me that one might take my egg as a starting-point. One would not think off-hand that science had played much part in putting it before me, yet as soon as we begin to consider, it leads us straight to some good examples of applied science. The egg was probably boiled in an aluminium saucepan, and cheap aluminium is emphatically a gift of science. For aluminium is not a metal like gold, where the difficulty is to find rich ore ; aluminium is the third commonest element in the world and exists in vast quantities around us in clay and common rocks, but, of course, in the combined state. We might call clay and rocks aluminium ores, in fact. The job of science, then, is to take the aluminium out of the rocks— cheaply.

Aluminium was discovered just over a hundred years ago by a German chemist, Wöhler, but it was very difficult and expensive to prepare by his method, so that in 1854 one pound of aluminium cost £50 sterling. It was a laboratory curiosity on account of its lightness and of scientific interest to chemists. Then Napoleon III—not Napoleon the Great, of course— thought it would be a fine substance for breastplates for his soldiers, and, as the French authorities often did and do, with great success, turned to a great man of science and asked him to find some way of making it more cheaply. In a few years the chemist, Sainte-Claire Deville, by his researches reduced the price to less than a tenth of that previously prevailing, and improvements in his process were made, until in 1889 the price had been reduced to 16s. a pound. About that time, that is, forty years ago, the electrical process now in use was introduced. This requires cheap electricity, which is produced by dynamos, which themselves grew out of Faraday's discoveries. To turn the dynamos requires cheap power. The British

Aluminium Company has therefore established its works in Scotland at Kinlochleven and at Foyers, where water power is available to turn great water turbines and generate the electricity. Your saucepan, then, depends upon the science of hydrodynamics, which governs the design of water turbines ; upon the science of electro-magnetism, which tells us how to design dynamos ; and the science of physical chemistry, which tells us how to employ electricity to make aluminium from the earthy clayey stuff called bauxite, which is an oxide of aluminium.

A few days ago someone asked me how it was that milk burned in an aluminium saucepan. I said this was not remarkable, as aluminium conducts heat very well; the remarkable thing was that the aluminium saucepan did not burn. For aluminium has a great affinity for oxygen, in suitable circumstances, and combines with it, that is, burns, giving out great heat. For instance, if the aluminium is finely divided, which gives it a larger surface—for every cut produces fresh surface in the same amount of substance—and mixed with certain oxides, and if the mixture, which is called thermit, be then heated at one spot, the aluminium burns with oxygen, producing such heat that iron can be melted. In fact, this thermit, this aluminium mixture, is often used for welding tram rails. Why, then, does not your saucepan burn if strongly heated ? Well, owing to the fact that the surface is not nearly as large as it would be if the saucepan were powdered, the oxidation takes place comparatively slowly, and then, luckily, ceases, for once a thin coat of oxide has been formed on the surface, it acts like a coat of varnish and protects the aluminium from further attack of oxygen. This protecting coat of oxide gives the aluminium saucepan its dull lustre, which contrasts with the silvery appearance of fresh-cut aluminium. This also accounts for the fact that you must not wash aluminium pots with soda, which attacks the protecting coat.

Your egg has probably been boiled over a ring on a gas-stove, which at once suggests another way in which science

serves the modern household.   In the first place, in your stove
the gas is burnt with air, which is drawn in by holes in the
gas-pipe, working on the plan of the Bunsen burner, which is
known to everyone who has ever been in a laboratory of any
kind.   This burner was devised by the famous chemist Bunsen
in 1855 to give a hot smokeless flame in his new laboratories at
Heidelberg—a famous university town where I pursued my
studies many years ago.   This, then, is a direct incursion of the
laboratory into the home.   Further, gas itself, as burnt at the
present day, is a product which has been improved, and is
being improved, by extensive researches.   Incidentally, one of
the products which is left behind at the gas retort is coal-tar,
from which practically all our dyes and many of our drugs are
made.   As for the gas itself, the problems have completely
changed in the last thirty years or so.   When men who are now
middle-aged were boys, gas was mainly burnt to give light,
that is, to give a bright flame : to-day it is only burnt to give
heat, for in the case of the incandescent mantle the gas itself
burns with a pale blue flame, as on the gas stove, and produces
light merely by its heating effect on the mantle.   The composition of the gas had therefore been adapted to this end, and
to-day you pay for gas, not by the cubic foot, but by the heating
power, the therm, which is clearly the only fair way.   One of
the most ingenious machines which I have ever seen is that
invented by the great physicist, Professor C. V. Boys (who
earlier in his life made the most accurate measurement of the
weight of the earth), for writing down automatically the heating
value of gas from minute to minute, allowing for all the
variations in the condition of supply that can take place.   Such
instruments are used by the big gas companies to make a
continuous record of the merits of the gas which boils your
egg.

Suppose we now are sitting in our room at night, as you
are.   Many of us will be using electric light.   Now the electric
lamp, simple as it looks, has probably had more highly scientific
thought spent on it than anything else in the house, and at the

present time some of the most brilliant scientific brains of the
world are employed in the research laboratories of the great
lamp manufacturers. You might think that once it has been
shown that a wire could be made hot by electricity the lamp
problem was solved, and it only remained to shut wires up in
bulbs for the electric lamp to be made ready for the market.
Hardly. In the first place a hot wire gives out both light and
heat, and we don't want the heat. The hotter the wire, the
more light and the more heat, but science has shown that as we
raise the temperature the bigger the proportion of the energy
that goes in useful light, and the smaller the proportion in
wasted heat, so that the problem is to get the wire as hot as
possible. This has led to the use of special metals for the wire,
metals which melt only at very high temperatures, especially
tungsten, and to handle tungsten is a problem of great scientific
interest. It is never melted, but is converted from a fragile
bar of compressed powder into a strong tough wire by special
processes known as swaging, which have had to be specially
devised. We want the metal of the wire in a special state, so
that it does not change its internal structure when heated, and
this has led to prolonged studies of the crystal form of the
metal. Again, if the lamp is to have a long life, we must be
very careful what is shut up in that bulb with it. In ordinary
low candle-power lamps the bulb must be pumped out as
thoroughly as possible. Traces of gases left in the lamp, or,
for that matter, in the wireless valve, lead to all kinds of
trouble, and to get rid of the last traces of gas is a terrible
problem. In the ordinary way, they are left stuck to the glass
when the bulb is exhausted of air, and then come off when the
glass is heated during the use of the lamp, and spoil the
vacuum. A long series of researches, which would take hours
to describe, has shown us, not how to get rid of the last traces
of gas, but how to make it stick to the glass for good, which
serves our purpose just as well.

    In the so-called half-watt lamps, which are supposed to give
one candle power for every half-watt of power—and perhaps

very nearly do—a special gas is deliberately put into the lamp to enable us to heat the filament hotter. The gas does this by interfering with the evaporation of the hot metal, but it also brings in fresh problems by conveying the heat of the wire to the glass. This effect is diminished by coiling the wire into a tight spiral. Then it was found that the wires were liable to a kind of disease, which made the wire thin in some spots and thick in others, and led to its speedy destruction. Research proved that this was due to traces of water vapour—tiny traces, nothing that you could see as dampness—left in the bulb. The molecules of water, by a complicated action, took the tungsten from one part of the wire and put it down in another. Nowadays the manufacturer takes particular care that there is no mischievous trace of water vapour in the lamp.

The wire is only one part of the story. The glass, the sealing in of the wires which bring in the current, the supporting wires for the hot filament, all have been the subject of long researches, the results of which are included in every lamp you buy. Properly looked at, an electric lamp contains so much knowledge that you may call it an encyclopædia of physics.

We could spend hours inside the house looking at examples of the result of scientific research, but let us, to finish up, look at the house itself, let us look at the problems of the builder. Remarkable results have been reached during the last few years by the application of scientific method to the study of building problems. There is a building research station under the Department of Scientific and Industrial Research where, among many other problems, the question of decay of building material, both brick and stone, has been studied. It has been found there that two stones may not agree together—that is, that if they are put in contact in a building, one may decay rapidly, and also, in a brick house, it has been found that re-pointing the mortar often leads to a rapid decay of the brick. Research has shown why. For instance, if limestone and sandstone be put together on the face of a building, what happens is this : when the face of the stone is wetted by the rain, which soaks in a short distance, certain soluble salts, mainly calcium sulphate, which exist in the limestone are dissolved. This solution of limestone salts then passes into the sandstone, being sucked up by the small pores of the stone. When this stone dries, the salts crystallize and become solid again, and in doing so burst up the surface of the sandstone. Similarly, in the case of a brick wall, in the ordinary way the soluble salts in the brick come out through the mortar : you can sometimes see efflorescence on mortar, which then decays. If the salts are shut in by re-pointing with a hard, impervious mortar they burst the brick.

The repair of the House of Parliament is a case in which this kind of scientific research is going to be very valuable. Two stones which weather quite satisfactorily when separate, may weather very badly when put together. The question is : What stones can be used together ? In the old days we should have had to build a small House of Parliament, and wait then twenty or fifty years, to see what happened. Nowadays we can find suitable stones to marry by working with a few pounds of stone in the laboratory.

THE SCIENCE OF CLOTHING

Now let us consider the way in which science plays its part in furnishing us with the clothes we wear at the price we pay. Until quite recent times practically all matters concerned in the manufacture of cloth and fabrics of all kinds were left to the engineer and the craftsman. The craftsman is a man for whom we all have the greatest admiration and respect, but often his very skill acts in some way as a check upon development. He is a man who, as the result of his own experience and that of the generations before him, can carry out some particular process—guess the temperature of a piece of metal or tell, by just feeling it, if a thread is sticky enough—with marvellous precision, and can get round difficulties as they arise by a kind of instinct. As long as he can do this, manufacturers do not bother to inquire why difficulties arise or how the tricks for dodging them work : they take it for granted that sometimes everything goes wrong, and hope that old Bill will put it right, if not to-day, to-morrow. When, however, the necessity of cheap and rapid production comes along, and we have to leave the processes more and more to machines, we must find out, if we are to be successful, just what is happening, just what are the properties of the substance we are dealing with, so that we can arrange for the proper conditions and for a perfectly definite job for our machine to do, day after day.

To-day some textile manufacturing concerns—the successful ones—are finding that it is very advisable to get as much precise knowledge as possible about the scientific properties of the fibres with which they deal. There are in England research associations of manufacturers for cotton, for wool, and for silk, all of which receive money grants and other encouragement from the Department of Scientific and Industrial Research, and in addition many big firms keep up large private laboratories in which men who have been trained in scientific research in the Universities and technical colleges work in close touch with the practical men, who know what problems arise in the factory.

Let us consider for a moment what has to be done before raw cotton, wool, or flax becomes a piece of cloth ready for the market. After these substances have been suitably cleaned of stuff that we do not want, they each consist of a mass of very fine fibres, and in the end it is the properties of these single fibres that govern everything, just as in the end it is the properties of the individual man that make up the behaviour of the fabric of a nation. Each kind of fibre—cotton, wool, flax—has its characteristic appearance under the microscope which is one of the most important aids that science offers the manufacturer. For instance, cotton consists of a mass of little hairs or fibres, about one-thousandth of an inch thick, and from half an inch to two inches long. These fibres are not round but of an irregular flattened shape, and they have twists in them, which are very important. They are like strong, fairly smooth, flattened tubes full of a softer jelly, let us call it, in the hollow space, called the lumen, inside. Wool fibres also have a marrow, as it were, but they can at once be distinguished under the microscope by the fact that they are covered with a kind of scale, something like fish-scales in appearance, which makes them cling very tightly when twisted, and also gives a kind of minute scaliness to the finished thread. Silk fibres are very smooth, which makes them reflect the light and shine, and artificial silk fibres are even smoother, as we can understand when we come to mention their manufacture. This smoothness explains why silk stockings ladder while woollen ones do not—the silk threads, which are like bundles of long eels, slip over one another, while the woollen threads, which are more like snakes, say, do not.

Now the manufacturer has to take these tiny fibres of cotton and wool, and in the first place twist them together to make a long sliver, or rough thread ; then to thin this rough sliver into a proper yarn ; then to weave this yarn into a fabric ; then to treat it in various ways to give it a good finish. At some step dyeing has to be done. I have not, for lack of time, mentioned all the processes, which, of course, differ in wool and cotton, but what I want you to observe is that during the passage to the finished cloth the little fibres have a very rough time, for they are pulled, twisted, wetted, dried, and heated. What we want to know, if we are to understand the difficulties that arise, is just how the little fibres behave under this trying treatment. First, how strong they are, what pull they will stand. Secondly, what kind of elastic properties they have. When a fibre is pulled it stretches : if the pull is left on it will go on stretching for some time. When the pull is taken off it does not go back to its old length at once. How it behaves depends upon how long it is pulled, how hard it is pulled, how the pull is taken off, and also, among other things, on what kind of pulls it has had during its previous history. The fibre behaves in some ways like a weak spring soaked in thick treacle, which clogs its movement. Also, if it be bent rapidly to and fro it gets tired, like a piece of metal wire does, and loses its springiness. All these effects are present in a piece of cloth as it comes from the loom. Some fibres are, for instance, more strained than others, and the different threads will slowly recover at different rates and make the cloth warp, if nothing is done to stop this. Actually the cloth is treated in various ways, with hot water, with steam, and so on, at different steps of the finishing process, to get rid of these strains. Just as glass is annealed, to get rid of the strains, and prevent it possibly cracking, so cloth is, as it were, annealed to get rid of possible cockling. Heating a suit of clothes in front of the fire to take out its wrinkles is, in fact, making it easier for the fibres to recover their original form, and so is very close to annealing a glass or metal.

Now that we have seen, in a very general way, the nature of the industry, we can understand why, if we went round a textile research laboratory, we should find some men studying the effect of loading and unloading, or of twisting and untwisting, these single fibres, like fine hairs ; others studying the effect of heat or of moisture on them ; others examining them under microscopes. You must not think it is easy to find out even roughly how a fibre behaves, for it is a very complicated thing, consisting, as it does, of this peculiar jelly inside a complicated wall. For instance, it has been shown that a single cotton fibre—only a thousandth of an inch across, remember—shows in its walls rings of daily growth, just as a tree shows rings of yearly growth.

What I have said so far is a very small part of the story. Research has given the cotton industry one of its most important processes, that called mercerization. As long ago as 1844 an Englishman called Mercer discovered that cotton fibres soaked in caustic soda solution—which chemists call sodium hydroxide and cotton-finishers call lye—swell up, shorten, and become much stronger. It was not until fifty years later, however, that the process was introduced into industry. To-day mercerized cotton is seen on every side : for instance, those charming silk-finish shirtings with which our shirt-makers tempt us are made of mercerized cotton. The effect of the alkaline solution is to swell up the pulpy inside of each cotton fibre, so that instead of being a flattened, shrunken twisted thing it is plump and round. The fibre has to be stretched at some stage during the process, so that here again we have many conditions to investigate. The effect of mercerization is not only to strengthen the cotton—it is the rare case of a chemical treatment which improves the quality of a fabric—but also to make it lustrous, like silk : the smooth surface of the plump fibres reflects light well. It also makes dyeing easier. If we were going through a textile laboratory we might perhaps see a man investigating the lustre produced in a given cotton by mercerization, by reflecting the light from it at a certain angle, and measuring the strength of the light by a photo-electric cell—a kind of electric eye. The scientific methods of measuring lustre are much more accurate than the judgment of the eye. Such a method would not be used to test the pieces of cloth that are being sold, but to find out what part in the process is really most effective in producing the lustre. There has been quite a lot of discussion among men of science as to what it is that makes the mercerized fibre lustrous. It seems settled now that it is due to a single reflection of light at the smooth surface of the plump fibre, just as if the fibre were a polished silver rod or wire. The lustre can be increased by the weaver arranging a number of mercerized fibres parallel to one another.

One of the most important contributions of science to our clothing—chiefly the clothing of the ladies—is the creation of an artificial fibre. Silk, wool, and cotton, are all produced inside the wonderful laboratory of the living organism—the living silkworm, the living sheep, or the living cotton plant. Artificial silk, however, is a fibre made by man, and the first which he has successfully produced. The silkworm produces a fibre by squirting from a gland a sticky liquid called fibroin which, on its way out, passes through a reservoir where it is surrounded by a gummy stuff called sericin. The worm actually produces two threads, whose gummy coatings stick together to form one. The great trick of the silkworm is that the fibroin thread hardens as soon as it comes out into a tough thread, the single silk fibre. The layer of sericin, which hardens more slowly, is washed off in the process of manufacture.

To imitate the worm we have to find some stuff which will form a smooth shiny thread ; we require a pulpy mass which can be squirted through small holes into threads, and which can be hardened as soon as the thread is formed. Science has shown that cellulose is a substance contained in all plant-cells. Something is known of its chemical structure, but not everything, so there is still something left to puzzle our research workers. It can be prepared from all kinds of plant substance, but the most usual source for ordinary cellulose is wood, although the finest kind of cellulose for photographic film is made from cotton. While we are talking of it, I may say that it is one of the most widely used materials of our civilization. Most paper is made from it, explosives are made from it, celluloid, artificial leather surfaces, varnishes, and dozens of other things : but we can speak only of artificial silk, and even of that we have only a word to say.

There are several ways of making a smooth shiny fibre from cellulose. One is to dissolve cellulose in a liquid made by putting copper oxide into ammonia. This makes a tough mass which can be squirted through fine holes into an acid bath, which takes away the ammonia solution again, and leaves us with the celluloid thread. This is what the Germans call " shine-stuff "—*Glanzstoff*. Another way is that invented by Cross and Bevan, who devoted their lives to the scientific investigation of cellulose. By treating the cellulose with alcohol and carbon disulphide they get a gelatinous stuff called cellulose xanthogenate, or viscose in the trade, which is squirted through fine holes into a bath which contains chemicals which turn it back to cellulose. If the viscose is squirted through a fine slit we get cellophane, as the shiny stuff used to wrap up chocolate-boxes and so on is called. It is really artificial silk in sheet form. In both these processes we add something to the cellulose to make it squirtable, and then take it back when the squirting is over, so as to leave a thread of pure cellulose. It is like giving a boy a shilling to crawl through a hole, and then taking the shilling back when he has done so. There is another process still, invented by the brothers Dreyfuss, which consists in what chemists call acetylating the cellulose, and thus leads to threads not of pure cellulose but of cellulose acetate. This is called Celanese. Whichever process is used, it is laboratory work that has shown how to turn trees into artificial silk stockings. The purified cellulose with which the manufacturer starts looks very like shiny white cardboard, and the final product is a beautiful thread.

So far we have not said a word about dyeing. In this respect science enters so deeply into our daily life that we cannot pass it by without a word, yet time will only allow us to spend a minute or two on the subject. Nearly all the dyes used to-day are prepared from coal-tar, which seventy years or so ago was considered to be so much rubbish. Science enters at every stage of their manufacture, right up to their testing for fading in light by means of ultra-violet rays. To make the dye stick to the fabric provides another series of problems on which science is perpetually engaged.

Let us consider now a girl going out to play tennis. She may be clothed with artificial silk, made by science from wood, that is from trees. The dyes with which she is coloured are made by science from coal formed from tree-ferns that have been buried for thousands of years. She carries, perhaps, a book, the paper of which is made from wood by processes worked out in the laboratory : a bag with a clasp also made from cellulose (or possibly from skim milk), and the soles of her shoes are made from a tree juice, rubber, by a process controlled at every step by science. We may therefore say that science is a good fairy who has taken a tree, or rather a few new wood splinters and a few very old wood splinters, waved her wand and produced the delicate clothing.

THE SCIENCE OF POWER

In conclusion, I want to glance very briefly at the service science renders in improving our sources of power.

Coal is by far the most important fuel which is won in Britain. Until comparatively recent times it was so cheap and plentiful that it was burnt anyhow, without much attention being paid to the smoke and dirt that occur with this haphazard handling. Our supplies of coal are, however, not unlimited nor is coal as cheap as it was at the time when our industries were built up. There is also a general feeling that smoke and dirt which can be avoided are bad things. Science has therefore been recently called in to work out how our coal can be used most advantageously. A lump of coal completely burnt will produce a certain amount of heat, but it is very important that, if it is burnt in a furnace, it should be *completely* burnt, that it should be burnt quickly, and that it should not form clinkers.

Another side of the coal problem is what is called carbonization. That is, the coal is heated in retorts, producing coke, the so-called coal-tar, and gas. About 40 million tons of coal are carbonized in England each year, and the question to be settled scientifically is, what is the best way of carrying out the carbonization when all three classes of products—coke, tar, and gas—are considered, more especially what is the best temperature for carbonization, or, as a cook would say, is it best, taking into account meat, gravy, and smell (that is, coke, tar, and gas), to cook our coal well done or underdone ? Are we to use high-temperature carbonization or low-temperature carbonization ? In general, carbonization gets over the smoke problem, for the two forms of fuel which it produces, coke and gas, burn without smoke.

We may, perhaps, end our talk by a glance at some of the more remote possibilities of power generation which have been considered by science. An enormous amount of energy is represented by the radiant heat of the sun, by the tides, and by atmospheric electricity. The trouble in all these cases is that whereas with water power of rivers and streams we have the energy comfortably localized, so that we can lay our hands on it without spreading, as it were, a net all over the countryside, with these other sources of power the energy, great as it is, is spread very thin. Even in Egypt, for instance, where the atmosphere is very dry and clear, only 60 horse-power has been generated by heat-collecting plant covering about an acre. This plant consists of long mirrors which focus the heat of the sun on special boilers and so generate low-pressure steam. So far very little has been done with the tides, and nothing with atmospheric electricity, and there does not seem much prospect of anything very sensational in either direction.

A possibility is the heat of the earth. The temperature rises as we go down a mine shaft : at some places the rise is as rapid as one degree Fahrenheit for every forty feet. In certain volcanic regions, in fact, for example in Central Tuscany, powerful jets of steam issue from cracks in the earth and the heat is used to drive turbines which generate electricity. In 1919 three large plants operating on this system were installed. In any case, the temperature must be very high a few miles down, and Sir Charles Parsons some years ago suggested as a possibility the sinking of a shaft some miles deep, which would not only give us a great deal of information as to the nature of the earth's crust, but also provide us with a practically inexhaustible supply of heat, were it found possible to install boilers at that depth and devise plant to make use of the steam.

From time to time we hear a great deal of the possibility that we shall one day learn to make use of the energy of the atom. In ordinary cases of burning fuel we are producing slight changes in the outer parts of countless millions of millions of atoms, but this is not what is meant. The atom itself has a structure : right at the centre of each atom there is a little compact structure called the nucleus, a kind of atom within an atom. We know from calculation that certain changes in the constitution of this nucleus would be accompanied by the release of relatively enormous quantities of energy. Now the nucleus, being surrounded by a kind of bodyguard of electrons, is beyond the reach of any of our ordinary laboratory methods of attack : raising a substance to a high pressure or a high temperature affects the outer parts of the atom, but does not produce any change in the nucleus. However, by the use of radium and similar substances, we can make an occasional nucleus give up a part of its energy, an effect which we can detect by very special methods. We have never obtained enough energy in this way for the amount of heat whose release has been deliberately provoked to be measurable, even by our most delicate apparatus. If, however, we could so arrange things that millions of millions of millions and millions of nuclei—which represent quite a small weight of stuff—should quickly undergo change, we should have one very great supply of power. If, to take the simplest case, we could make, by nuclear transformation, an ounce of hydrogen turn into helium we should have a million horse-power for a seven-hour day. . . .

This concludes my talk. I have tried to give you some idea of the way in which science is at work all round you, investigating the most ordinary things, from threads and saucepans to bricks and lumps of coal, in an endeavour to make the world an easier place to live in for the many people who to-day inhabit it. I hope, even if I have not made everything clear, that I may at least have left some impression of the ceaseless self-questioning, the honest search, by which science lives and grows.

" All this is true " some will argue " but think how much more horrible wars are to-day, thanks to science." Wars are always horrible, and we certainly cannot blame the scientists for starting the wars. *Is science more a Preserver than a Destroyer ?* The following lecture was given in the dark days of the first World War, and in it Sir William Osler, one of the greatest doctors of his age, tells us that " the wounded soldier would throw his sword into the scale for science—and he is right ".

SIR WILLIAM OSLER

SCIENCE, WAR, AND PEACE

IN Time our civilization is but a thin fringe like the layer of living polyps on the coral reef, capping the dead generations on which it rests. The lust of war is still in the blood : we cannot help it. There was, and there is as yet, no final appeal but to the ordeal of battle. Only let us get the race in its true perspective in which a thousand years are but as yesterday, and in which we are contemporaries of the Babylonians and Egyptians and all together within Plato's year. Let us remember, too, that war is a human development, unknown to other animals. Though Nature is ruthless " in tooth and claw ", collective war between members of the same species is not one of her weapons ; and in this sense Hobbes' dictum that " war was a state of Nature " is not true. The dinosaurs and pterodactyls and the mastodons did not perish in a struggle for existence against members of their own species, but were losers in a battle against conditions of Nature which others found possible to overcome. In our own day the gradual disappearance of native populations is due as much to whisky and disease as to powder and shot, as witness in illustration of the one the North American Indian and of the other the Tasmanians.

Some of us had indulged the fond hope that in the power man had gained over Nature had arisen possibilities for intellectual and social development such as to control collectively his morals and emotions, so that the nations would not learn war any more. We were foolish enough to think that where Christianity had failed science might succeed, forgetting that the hopelessness of the failure of the Gospel lay not in the message, but in its interpretation. The promised peace was for the individual—the world was to have tribulations ; and Christ expressly said : " Think not that I am come to send peace on earth ; I came not to send peace but a sword ". The Abou Ben Adhems woke daily from their dreams of peace, and lectured and published pamphlets and held congresses, while Krupp built 17-inch howitzers and the gun range of the super-Dreadnoughts increased to eighteen miles !

And we had become so polite and civil, so cultured in both senses of that horrid word, with an " Is thy servant a dog ? " attitude of mind in which we overlooked the fact that beneath a skin-deep civilization were the same old elemental passions ready to burst forth.

In spite of unspeakable horrors war has been one of the master forces in the evolution of a race of beings that has taken several millions of years to reach its present position. During a brief fragment of this time—ten thousand or more years—certain communities have become civilized, as we say, without however, losing the savage instincts ground into the very fibre of their being by long ages of conflict. Suddenly, within a few generations, man finds himself master of the forces of Nature. In the fullness of time a new dispensation has come into the world. Let us see in what way it has influenced his oldest and most attractive occupation.

Science is a way of looking at the world taught us by the Greeks—a study of Nature with a view of utilizing her forces in the service of man. It " arose from the simplest facts of common experience, and grew by the co-operation of the mass of men with human intellect at its highest. And when developed it returns again to strengthen the common intelligence and increase the common good. Above all, more perfectly than any other form of thought, it embodies the union of past and present in a conscious and active force ". Man's latest acquisition, it has worked a revolution in every aspect of his life, without so far changing in any way his nature. He is still a bit bewildered, and not quite certain whether or not the invention is a Frankenstein monster. The promise of Eden of full dominion over Nature has only been fulfilled in our day. The flower and fruitage has come suddenly within a couple of generations. Even the seed-time was but a few years ago, for to the Heidelberg man, looking down the ages from the Glacial period, Aristotle and Darwin are contemporaries, Galen and Lister fellow-practitioners. Steam and electricity have upset out week-day relations, and the theory of evolution our Sundays. Like a beggar suddenly enriched, man has not yet found himself; and the old ways and old conditions often sort ill with the changing times. New bottles could not always be found for the new wine.

Organized knowledge, science, if living, must infiltrate every activity of human life. There was a difficulty in these islands, which of fruitful ideas, inventions and discoveries have had the lion's share, but failed to grasp quickly their practical importance. The leaders of intellectual and political thought were not awake when the dawn appeared. The oligarchy who ruled politically were ignorant, the hierarchy who ruled intellectually were hostile. Read of the struggles at Oxford and Cambridge in the 'fifties and 'sixties of the last century to get an idea of the attitude of the intellectual leaders of the country towards " Stinks ", the generic term for science. It was not port and prejudice, as in Gibbon's day, but just the hostility of pure mediæval ignorance. Those in control of education were more concerned with the issues of Tract 90 and the Colenso case than the conservation of energy and *The Origin of Species*. To take but one example. What a change it might have wrought in rural England, if in 1840, when the distinguished Professor Daubeny was made professor of rural economy, Oxford could have had great State endowment for an Agricultural College. The seed was abundant and the soil was good, and only needed the cultivation that has been given so freely by members of the past generation, with what results we see to-day at Oxford and Cambridge and in the new universities.

In two ways science is the best friend war has ever had; it has made slaughter possible on a scale never dreamt of before, and it has enormously increased man's capacity to maim and to disable his fellow-man. In exploiting the peaceful victories of Minerva, Mars has added new glories to his name. More men are killed, more men are wounded, and consequently more men are needed than ever before in the history of the world's wars. From 1790 to 1913 there were 18,552,200 men engaged in the great wars, of whom 5,498,097 lost their lives (D. E. Smith). In the Balkan wars of 1912-13 there were 1,230,000 men engaged, of whom 350,000 were killed. In the Russo-Japanese War there were 2,500,000 men, of whom 55,900 lost their lives (D. E. Smith). It is estimated that in the present war more than 21 millions are engaged! As weapons have improved the losses will be yet greater, and we may expect that at least 5 or 6 millions of men in the prime of life will be killed. Within a few years artillery and high explosives, submarines, and aircraft have so revolutionized our methods of warfare that thousands are now destroyed instead of hundreds. The rifle and the bayonet seem antiquated, and one may go from hospital to hospital and find not a wound from the latter, and comparatively few from the former.

Let us see what science has done in a mission of salvation amid the horrors of war. Through the bitter experiences of the Napoleonic wars, of the Crimea, of the American Civil War, and more particularly of the recent campaigns, there has been evolved a wonderful machinery, replete with science, for the transport and care of the sick and wounded. There must be suffering—that is war—but let us be thankful for its reduction to a minimum, through the application in every direction of mechanical and other pain-saving devices.

If the foes of our own household, the " anti's ", would spend a few days at a hospital for infectious diseases, see the modern methods, and learn a few elementary facts about immunity, they could not but be impressed with the applications of scientific horticulture to disease, and be lost in admiration of a technique of extraordinary simplicity and accuracy.

The second great victory of science in war is the prevention of disease. Apollo, the " far darter ", is a greater foe to man than Mars. " War slays its thousands, Peace its ten thousands." In the Punjab alone, in twelve years, plague has killed 2½ millions of our fellow-citizens. This year two preventable diseases will destroy more people in this land than the Germans. The tubercle bacillus alone will kill more in Leeds in 1915 than the city will lose of its men in battle. Pestilence has always dogged the footsteps of war, and the saying is true— " Disease, not battle, digs the soldiers' grave ". Bacilli and bullets have been as David and Saul, and at the breath of fever whole armies have melted away, even before they have reached the field. The fates of campaigns have been decided by mosquitoes and flies. The death of a soldier from disease merits the reproach of Armstrong :—

" Her bravest sons keen for the fight have died
    The death of cowards and of common men—
    Sunk void of wounds and fall'n without renown."

This reproach science has wiped away. Forty years ago we did not know the cause of any of the great infections. Patient study in many lands has unlocked their secrets. Of all the great camp diseases—plague, cholera, malaria, yellow fever, typhoid fever, typhus, and dysentery—we know the mode of transmission, and of all but yellow fever the germs. Man has now control of the most malign of Nature's forces in a way never dreamt of by our fathers. A study of her laws, an observation of her facts—often of very simple facts—has put us in possession of life-saving powers nothing short of miraculous. The old experimental method, combined with the new chemistry applied to disease, has opened a glorious chapter in man's history. Half a century has done more than a hundred centuries to solve the problem of the first importance in his progress.

Lastly, in the treatment of wounds science has made great advances. The recognition by Lister of the relation of germs to suppuration, an outcome of Pasteur's work, has done away with sepsis in civil life. High explosives, shell, and shrapnel make wounds that are at once infected by the clothing and dirt, and are almost impossible to sterilize by any means at our command, but with free drainage, promotion of natural lavage from the tissues by Wright's method, and the use of antiseptics when indicated, even the most formidable injuries do well. The terrible laceration of soft parts and bones adds enormously to the difficulty of treatment. The X-ray has proved a boon for which surgery cannot be too grateful to Röntgen and to the scores of diligent workers who have given us a technique of remarkable accuracy. Other electrical means for detecting foreign bodies have also good results.

Of the germs blown into wounds from the soil and clothing and skin the pus-formers are the most numerous and most important. Two others have proved serious foes in this war, the germ that causes gas gangrene and the tetanus bacillus. I am told that methods of treatment of wounds infected by the former are giving increasingly good results. The soil upon which the fighting has occurred in France and Flanders is rich in the spores of the tetanus bacillus ; the disease caused by it was at first very common and terribly fatal among the wounded. For centuries it has been one of the most dreaded of human maladies, and justly so, as it is second to none in fatality and in the painful severity of the symptoms. No single aspect of preventive medicine has been more gratifying in this war than the practical stamping out of the disease by preventive inoculation. In the first six months of this year only thirty-six of those who were inoculated within twenty-four hours of being wounded suffered from tetanus.

And what shall be our final judgment—for or against science ? War is more terrible, more devastating, more brutal in its butchery, and the organization of the forces of Nature has enabled man to wage it on a titanic scale. More men will be engaged and more will be killed and wounded in a couple of years than in the wars of the previous century. To humanity in the gross science seems a monster, but on the other side is a great credit balance—the enormous number spared the misery of sickness, the unspeakable tortures saved by anæsthesia, the more prompt care of the wounded, the better surgical technique, the lessened time in convalescence, the whole organization of nursing ; the wounded soldier would throw his sword into the scale for science—and he is right.

The fact that science, on balance, is of great value to mankind is brought home even more vividly to us if we take a wider view, and see how it affects, or may affect, life over a whole continent. *How has science helped Africa : and how can Africa help science ?* The answer is given in Jan Hofmeyr's presidential address to the British Association when it met in Capetown in 1929. Hofmeyr was scholar and scientist as well as statesman, and the breadth of his interests can be gauged from his speech.

JAN H. HOFMEYR

AFRICA AND SCIENCE

TO-NIGHT I enter upon the consummation of what is at once the highest and the least merited distinction which it has been my privilege to receive. To those who called me to the office of President of the South African Association for the Advancement of Science I tender my sincere thanks. I make myself no illusions in respect of the adequacy of my claims to that honour on the ground cither of scientific attainment or of services rendered to the cause of Science, nor would I have our visitors remain for a moment without the knowledge that my scientific qualifications for this Presidential Chair are of the slightest. They are far less, indeed, than those of that distinguished statesman to whom, when he had remarked to the great Faraday in relation to an important new discovery in Science, " But after all, what is the use of it ? " the scientist replied, " Why, sir, there is every probability that you will soon be able to tax it ". The Presidency of this Association is an honour the conferment of which upon myself has never seemed to fall properly within the scope of my ambitions ; it imposes responsibilities for the discharge of which I am all too scantily equipped ; and I can only seek to justify my election in a manner similar to that which Mr. Stanley Baldwin followed when he was chosen to be President of the Classical Association in England. I can but say that, while it is to the scientist that we look for the advancement and the progress of Science, the effectiveness with which his work is brought to fruition does depend in some measure on the interest, the sympathy, and the enthusiasm with which his achievements are followed up by that army of plain, ordinary men, in which I gladly count myself a musket bearer. In no other capacity dare I venture to address you. It was once said by a literary man of some distinction that the man of science appears to be the only man in the world who has something to say, and he is the only man who does not know how to say it. There is an obvious rejoinder, that the man of letters frequently has nothing to say, but says it at great length. I dare not claim to be a man of science. I can only hope that I shall not be deemed to-night to have qualified for consideration as a man of letters in the sense of that retort. . . .

The honour which has been conferred upon me is the greater because of the special significance which attaches to my year of office. It is the year of the keenly anticipated second visit of the British Association for the Advancement of Science to South Africa, and for that reason my first words to-night are, happily, words of welcome. Not merely the Association for which I speak, but all South Africa, rejoices in the presence of the British Association and its distinguished members. To its parent body, which can look back upon all but a century of glorious achievement, this stripling Association brings its tribute of respectful admiration and good-will. . . .

If South Africa aspires to leadership in Africa in other branches of activity, why not also in Science ? If the outlook of the nation is broadening, why should not its scientists also begin to think in continents ? If as a people we are anxious to make our contribution to Africa, eager to give it of our best, rather than to get from it that which will be to our material advantage, why should not our Science also become consciously and deliberately African in its outlook, its ideals, and the tasks to which it applies itself. If Science has consolidated its position in South Africa, as we believe it has, is it not fitting that, with South Africa as its base, it should enter now into the new sphere of opportunity and achievement which stretches mightily outwards from its borders ?

To you, our visitors, I look to give us the stimulus and the encouragement to that enterprise. You have come to Africa. This great land-mass which has reared itself against time's passage, almost since time's beginning, and holds inviolate so many of the records of that passage, has challenged your attention. You have come to Africa to seek new inspiration for the study of the problems that interest you, by seeing them against a different background which has for many of you an unaccustomed vastness. But while Africa was your goal, you did not think fit to enter it at the point nearest to your homes. You steamed down, day after day, skirting the long coast-line of this vast expanse of veld and forest, and have entered it by its Southern gateway. For a great body of scientists, it is the only point of effective entry into Africa. It is by way of this Southern gateway that Science itself can most effectively be made to permeate Africa. And to you, having so come, to you, the ambassadors of Science, I present—Africa. It is Africa and Science which, I would like to think, are to-day met together. Happy indeed should be the fruits of the mating.

It is to that theme—Africa and Science—that I propose now to invite your attention. What can Africa give to Science ? What can Science give to Africa ? Those are the questions to which I would address myself. But as I speak, I would ask you all to remember that it is for the South African scientist that the answers to these questions have primary significance. It is for him that they have significance, because, for the solution of many of the problems of South Africa a greater knowledge of Africa as a whole than is at present available is essential, and the extension of that knowledge is his personal responsibility. It is for him that they have significance because he dwells in a land which is strategically placed for attacking the problems of Africa and for drawing forth its hidden resources of scientific discovery for the enrichment of Science throughout the world.

What, then, can Africa give to Science? In reply to that I can do no more than suggest some of the lines along which Africa seems to be called upon to make a distinctive contribution to Science.

First there are the related fields of Astronomy and Meteorology. To Astronomy I shall but make a passing reference. This continent of Africa, more especially the highlands of its interior plateau, with its clear skies and its cloudless nights, offers wonderful facilities to the astronomer. As proof of the necessity of utilizing those facilities, especially with a view to the study of the Southern heavens, I need but quote the words used by Professor Kapteyn on the occasion of the 1905 visit :

> " In all researches bearing on the construction of the universe of stars, the investigator is hindered by our ignorance of the Southern heavens. Work is accumulating in the North, which is to a great extent useless until similar work is done in the South."

Africa has to its credit considerable achievements in the past in the field of astronomical research. The increased equipment now available should make it possible to increase greatly the amount of systematic work now being done, and to offer important contributions to astronomical science.

But probably of greater importance is the work waiting to be done in Meteorology. Few branches of Science have a more direct effect upon the welfare of mankind—that is a lesson which we in South Africa should have learnt only too well—but in few has less progress been made. And in meteorological work Africa is probably the most backward of the continents. It is not so long since Dr. Simpson of the London Meteorological Office declared that, save from Egypt, his office received practically no meteorological information from the great continent of Africa. Moreover, the backwardness of Meteorology is in large measure due to the intricacy of the problems involved, and the necessity of having world-wide information made available. The problems of Meteorology are emphatically not the problems of one country or of one region. The South African meteorologist must see his problems *sub specie Africæ* (the seasonal changes in South Africa depend on the northward and southward oscillations of the great atmospheric system overlying the continent as a whole); and quite apart from what he can learn from the rest of Africa, the Antarctic regions have much to teach him. But while the development of meteorological research throughout Africa is of supreme economic importance for Africa, Africa in its turn has its contributions to make to other continents. In particular should we not forget the close interrelation of the meteorological problems of the lands of the Southern hemisphere. The central position of Africa in relation to those lands gives not only special opportunities but also special responsibilities for meteorological observation and research. For the sake both of South Africa and of Science in general I would venture to express the hope that this second visit of the British Association will give as powerful a stimulus to Meteorology as did the first to Astronomy.

Next, I would refer to Africa's potential contributions to Geological Science. Africa is a continent, portions of which have always had a special interest for the geologist because of the great diversity of the geological phenomena manifested, and the vast mineral wealth which, as its ancient workings so abundantly prove, has attracted man's industry from the very earliest times. But in our day the opportunities which it offers to the geologist to make contributions to the wider problems of Science are coming to be more fully realized than ever before. Of special interest in this connexion is the light which African Geology, more especially in the form of the study of ancient glacial deposits, can throw on the Wegener hypothesis of continental drift. In the past our geologists have thought mainly of the correlation of our formations with those of Europe. It is time that they paid more attention to their possible affiliations with those of the continents to east and west of us. If Geology can establish the hypothesis that Africa is the mother continent from which India, Madagascar, and Australia on the one side and South America on the other have been dislodged, it will give a new orientation to many branches of scientific activity. For that investigation also Africa occupies a central and determinative position in relation to the other continents, such as we have noted to be the case in the sphere of Meteorology. There are many other geological problems on which Africa can probably shed much light. There is, for instance, the constitution of the earth's deeper sub-strata, in regard to which, as Dr. Wagner has recently pointed out, the study of the volcanic Kimberlite pipes, so numerous throughout Africa south of the Equator, and of the xenoliths they contain, including the determination of their radium and thorium contents, may be of the greatest significance. There is the possibility that the exploitation of Africa's great wealth in potentially fossil-bearing rocks of presumably pre-Cambrian age will yet yield us remains of living beings more primitive than any yet discovered ; there are the great opportunities of study which the African deserts offer in the field of desert Geology and Morphology, and there is the assistance which African Geology has rendered to vertebrate and plant Palæontology, and can render to African Anthropology in the investigation of this great museum of human remains and relics, which we call the continent of Africa.

I pass on to Medical Science. I have referred already to

the contributions to the study of the problems of industrial medicine and hygiene which the special circumstances of the South African gold-mining industry have made possible. Those contributions have, we may well hope, but prepared the way for advances of a revolutionary character in the early detection, prevention, and treatment of all forms of respiratory disease. But even greater are the opportunities which the continent of Africa offers for the study of tropical diseases, of which it may well be described as the homeland. In Africa there have been and necessarily must be studied the problems connected with malaria, blackwater fever, sleeping sickness, yellow fever, and many other scourges of civilization, and from Africa there may well come hope and healing for mankind. There are other problems of Medical Science for the study of which Africa is uniquely fitted. There are the physiological questions, important also from the political point of view, which bear on the fitness of the white races to maintain a healthy existence in tropical surroundings, at high altitudes, and in excessive sunlight. For these investigations the diversity of conditions prevailing in the various regions of the African continent make it a magnificent natural laboratory. There is the elucidation of the factors which account for the varying susceptibility of white and coloured races to acute infectious diseases, tuberculosis, and certain types of malignant disease, together with the light which such elucidation may throw on the physical and chemical composition of the human body. Lastly, I would mention the exploration of that most interesting borderland between Psychiatry and Psychological Science by an analysis of the mentality of the diverse African peoples. That investigation has an important bearing not only on the limitations and capacities of racial intelligence, but also on the methods which the ruling races in Africa should follow in seeking to discharge their obligations towards their uncivilized and unenlightened fellow-Africans.

Closely linked with Medical Science is the study of Animal Biology. In some instances the problems of the two branches of Science are to be approached along parallel lines ; in others biological investigations are fundamental to the growth of Medical Science ; of no less significance is that unity which there is in Nature, making it possible for the truths of Animal Biology to be translated into facts concerning mankind. In the African continent there is no lack of opportunity to advance Science by physiological inquiries into animal structure, by the isolation of the parasites of human and animal diseases, and by the tracing of the life histories more especially of the minuter forms of animal life. " Nowadays " in the words of Professor J. A. Thomson " the serpent that bites man's heel is in nine cases out of ten microscopic." But scarcely less important are the extensive facilities which Africa still offers for the study of the habits and behaviour of the larger mammals. The naturalistic study of these animals, not as stuffed museum species, but in the laboratories of their native environment, has received all too scanty attention from the scientist, and this is a reproach which African Science, with its rich dowry of mammal and primate material, may confidently be expected to remove. Nor will this study of animal behaviour, especially of those animals which approach nearest to the human type, be without its bearings on our investigations of the workings of the human mind. If in this hasty survey I may take time to mention one more point within this field, I would refer to the results which await the intensified activity of the marine biologist and the oceanographer in the as yet all but virgin territory of the African coast-line. This Association of ours has long dreamed of an African Marine Biological Station as broad in its conception and comparably as useful from the wider scientific and the more narrowly economic points of view as those of Plymouth or Naples or Woods Hole, and withal a rallying-point for the naturalist, the zoologist, the botanist, the geographer, the anatomist, the physiologist, indeed for all those workers whose diverse problems meet at the margin of the sea.

From Animal Biology we pass by an easy transition to

Anthropology, the study of man himself. And here Africa seems full of splendid promise of discovery that may verify Darwin's belief in the probability that somewhere in this land-mass was the scene of Nature's greatest creative effort. It would seem to be not without significance that Africa possesses in the chimpanzee and the gorilla those primate types which approach most nearly the form and structure of primitive man. To that must be added that in the Bushman, Pygmy, and negroid races Africa has at least two and possibly three early human stocks which are characteristically her own and belong to no other continent. No less striking is the fact that in Gibraltar, in Malta, and in Palestine, that is, at each and every one of the three portals into Africa from Europe and Asia in Pleistocene times, there have been discovered evidences of the presence of Neanderthal man. In Africa itself there was found at Broken Hill some nine years ago a skull with the most primitive or bestial facial form yet seen, and so closely akin to the Neanderthal stock as to establish firmly the expectation of finding further compelling evidence of a long-continued Neanderthaloid occupation of the African continent. The discovery at Taungs on the one hand, which reaches out to-wards the unknown past, and the finds at Boskop and in the Tsitsikama on the other, which assist in linking up the period of Rhodesian man with the coming of the Bushfolk, open up to us, in conjunction with the aforementioned facts, a vista of anthropological continuity in Africa such as no other continent can offer. The recent investigations in the Great Rift Valley, near Elementeita in Kenya, and the fossil discoveries on the Springbok Flats, north of Pretoria, have again fixed the atten-tion of the anthropologist on Africa.

Nor are the data presently available restricted to these discoveries. The efforts of archæologists, and the application of improved scientific methods in excavation, are giving us stratigraphical evidence of the succession of stone cultures which is of the utmost importance. I have already men-tioned the assistance which Geology can render in this work, but there is needed also the co-operation of those who labour in the converging fields of Anatomy, Archæology, Palæontology, and Comparative Zoology. That co-operation has already commenced. In the investigation of the Vaal River gravels it has yielded important results, and we may look forward to its continuance and expansion in the years that lie ahead. Of the importance of African Anthropology for the understanding of that of Europe there can be no question. Work of importance has already been done in the study of the relations between Palæolithic Art in Europe and Palæolithic Art in Africa. The significance of these comparisons is but emblematic of the importance of similar investigations in regard to stone cultures, rock engravings, ancient mining, stone circles and ancient ruins, methods of primitive mining and agriculture, tribal organization, laws and customs, indeed the whole range of the hitherto unexplained or partially explained phenomena of living and extinct cultures. There is no lack of avenues which the student of African Anthropology may follow in the hope of finding at the end of them results of supreme value for Science in general.

I would speak next of the vast field, as yet almost uncharted, of phonological and philological study. Here in Africa we have great opportunities for the examination of linguistic problems, and some of them have bearings which extend far beyond the limits of Africa. One thinks first of the opportunities which Africa offers for investigating the results of the transplantation of languages, which have a long history of cultural development behind them, to regions inhabited by primitive peoples. Here there are two sets of phenomena, each with its own special interest. On the one hand we have the modification of the languages of those European peoples, who have established themselves in Africa as permanent settled communities, under pressure of the new linguistic influences into contact with which they have been brought. Of these phenomena the study of Afrikaans offers perhaps the best examples to be found in the whole field of linguistics—

its importance for the student of comparative philology is very far from being adequately appreciated. On the other hand we have those cases where European languages have come to Africa as the languages not of settled communities, but of officials and others like them who are but temporarily domiciled in this continent, and leave no descendants behind them to carry on the process of evolution of distinctive forms of speech. Here the phenomena which are of interest to the student of linguistics are to be found in the wealth of deformation and adaptation which the native populations have introduced in their endeavours to speak the European languages of their rulers. Work such as has been done by Schuchardt in Negro-Portuguese and Negro-French opens up a wide area of most attractive investigation.

But the most important task in the field of African linguistics is the actual recording the native languages of Africa, our backwardness in respect of which is a reproach to Science. Such study is, of course, important in relation to Africa itself, but of even greater significance for my present purpose is its bearing on scientific problems of wider scope. In that connexion I would suggest two points. We are still only at the beginning of the study of Comparative Bantu. That in due course should lead to a knowledge of Ur-Bantu. Such a study and such a knowledge will necessarily be of importance to the comparative philologist, both because of the light shed by the study of one group of languages on the study of other groups, and also because it opens the way to the investigation of the relationship of Bantu to the other African tongues, and its place in the general scheme of the languages of the world. But of even greater interest is the study of African languages as throwing light on the interpenetrations and interactions of primitive peoples. Language is a function of social relationship, and its study is therefore of great value for ethnological and historical investigations. May I give one instance of what I have in mind? Two millennia back South-West Arabia was the seat of the powerful commercial civilizations of the

Mineans, the Sabæans, and the Himyarites, radiating east-wards to India and south-westwards to Africa. The extent of their relationship with Africa it has hitherto been most difficult to trace, but linguistic evidence may prove to be of great value. Professor Maingard has pointed out to me that the Makaranga who live near Zimbabwe call water " Bahri ", a word closely related in form to " Bahr ", the " sea " of the Arabs, although the Makaranga themselves are not a sea-board people, and that " Shava " is their word for " to sell or barter ", while to the Himyarites " Saba " meant " to travel for a commercial pur-pose ". Not less suggestive is the study of place-names, and while I do not suggest that I have evidence on which any con-clusion can be based, I do contend that these investigations may prove to be of a most fruitful character. It would be interesting indeed to see what evidence linguistics can bring in respect of the relationship of South Africa with Madagascar, and also with Polynesia through Madagascar, where the tribe once dominant politically, the copper-coloured Hova, are ethnologically and linguistically Melanesians amid the darker-hued Sakalavas and other negroid tribes. It may even be that such studies will conjure up to our minds pictures of great migratory movements with Arab dhows and South Sea praos cleaving the waters of the Indian Ocean. Only last year a canoe constructed of wood native to South-Eastern Asia was found in Algoa Bay.

And, finally, in this survey of what Africa can give to Science I would refer, with the utmost brevity perforce, to Africa as a field, favoured as is no other, for the study of all those com-plicated problems which arise from the contact of races of different colours and at diverse stages of civilization. Of those problems, ranging from the investigations of the bio-logical factors involved in the conception of race to the prac-tical problems of the administration of backward peoples I need not speak. They have come to be part almost of the everyday thinking of most civilized men. What I would emphasize is that in Africa, as nowhere else, the factors which constitute these problems can be studied both in isolation and in varying degrees of complexity of interrelationship, that in Africa we have a great laboratory in which to-day there are going on before our eyes experiments which put to the test diverse social and political theories as to the relations between white and coloured races, that in Africa there are racial problems which demand solution, and the solution of which will affect or determine the handling of similar problems throughout the world. We hear men speak of the clash of colour, and are sometimes told that Africa is the strategic point in that struggle. I think of it rather as the continent which offers the richest opportunities to those who would investigate racial problems in the true spirit of Science, and so discover the solutions, which may yet enable that clash to be averted and the threat which it implies to our civilization to be dispelled.

I have sought—briefly and all too inadequately—to in-dicate some of the lines along which Africa seems to be able to make a distinctive contribution to Science. It remains for me, yet more briefly, to speak of Africa's challenge to Science, and to seek to answer the question : What can Science give to Africa ? I shall not stop to emphasize the point, that the greatness of Africa's potential contributions to Science, the key which perhaps she holds to the riddle of human origins, the intriguing vistas opened up in the study of her relationship with South America and Australasia with its suggestion of past continental continuity, that all these and more constitute a challenge to Science to actualize those potentialities. Let me seek rather to define the twofold challenge of Africa in another way. Firstly, Africa defies Science to unravel her past. Throughout history she has ever been the continent of mystery. She was so to that pioneer of geographers, Herodotus, to whom nothing that was told him about Africa was so im-probable that he declined to give it credence. She was so to the Romans, who regarded Africa as the natural home and source of what was strange and novel and unaccustomed.

She was so to the navigators who did so much to break down the barrier wall between the Middle Ages and the modern world. And though in our day the geographical mysteries of Africa have in large measure been solved, the work of the prober of her scientific secrets is only beginning. Then, secondly, Africa challenges Science to define, to determine, and to guide her future. If the great resources of this vast undeveloped continent are to be made available for humanity in our own and the succeeding generations, Science must make it possible for the man of European race to undertake that work of development, by showing him how to protect himself, his stock, and his crops against disease, by enabling him to conserve and utilize to the greatest extent the soils, the vegetation, and the water supplies of the continent, by bringing to bear the resources of modern engineering on the exploitation of its wealth, and not least by determining the lines along which white and coloured races can best live together in harmony and to their common advantage.

That is the challenge of Africa to civilization and to Science. It is not now thrown out for the first time ; it is not the first time that it will have been taken up. It is in Africa that the Greco-Roman civilization won some of its most glorious triumphs; in Africa that the spade of the archæologist has in our day, by uncovering great Roman towns with noble public buildings and efficient irrigation systems, provided impressive evidence of the magnitude of the achievement of Roman Imperialism. But Rome failed to conquer Africa for civilization, and left the challenge to those who were to follow after. She failed chiefly for two reasons : the might of African barbarism and the defiant resistance of African nature. We in our day, confronted by the same challenge, still have the same enemies, hitherto victorious, to contend against. But we meet them with the advantage of having resources at our disposal which our Roman predecessors lacked. It is to use those resources effectively that Africa challenges Science.

In dealing with African barbarism we have weapons such

as Rome could never dream of, and not the least valuable are those provided by the scientific investigation of the native peoples of Africa. The way to the solution of the problems presented by African barbarism is to be sought in an understanding of the character and mentality of primitive peoples, in the exploration of those regions in their social life where are to be found the factors that determine their reaction to diverse methods of administration. The study of African languages and of African anthropology is therefore fundamental to the development of the continent. For that work Africa possesses special advantages, and one can but hope that the facilities now being built up in our South African Universities will be recognized in Britain and elsewhere, and become an important factor in the response of Science to the challenge of Africa.

Not less formidable is the conquest of African nature, for the achievement of which also we in our day are far better placed than were the Romans. It is modern Science which gives us that advantage. Three great tasks confront Science in the conquest of African nature. First, Science must make Africa safe for the white man to live in. I have spoken of the opportunities which Africa offers for the study of tropical diseases as likely to yield results of significance for Science in general. But primarily will those results be of significance for the development of Africa? This part of the challenge of Africa is not lightly to be taken up. Africa has taken heavy toll of Science. The recent deaths in Nigeria of Stokes, Young, and Noguchi, worthy followers in the tradition of Lazear and Myers, are a reaffirmation of the gravity and insistence of that challenge. The importance for the cause of civilization of a successful response to that challenge cannot be illustrated better than by the story of the construction of the Panama Canal. De Lesseps attempted the task and failed. For every cubic yard of earth excavated by him a human life was sacrificed to yellow fever or malaria. It was the successful attack some twenty years later on the death-dealing mosquito under the direction of General Gorgas that made possible the completion of one of the most important engineering enterprises of modern times.

Secondly, Science must combat the foes which have to be contended with in the development of African agriculture. Africa is prodigal indeed in the production of insect and other foes to cattle and to crops. Science is already making an effective response to this part of the challenge. But there is much that remains to be done. And we shall be none the worse for the timely realization by the politician and the administrator of the contributions which Science can make. All too often in the past settlement schemes have been undertaken and ended in disaster in areas unhealthy to man, beast, or crops, when, if the scientist had first been called in, precautions might have been taken which would have averted the calamity.

Finally, Science must harness the great resources of Africa. And here there are suggested to us all the varied contributions which the engineer can make in the work of development. Has not the Institution of Civil Engineers defined the ideal underlying all engineering activity as " the art of directing the great sources of power in nature for the use and convenience of man " ? Africa offers abundance of opportunities for the realization of that ideal. It is not by working in isolation that the engineer will realize it, but rather by co-operation with his colleagues in other branches of Science, and by the correlation and co-ordination of the essential data which they must do so much to provide. First in the order of engineering development come the civil and mining engineers. Their tasks are the provision of facilities for communication, for health, for the conservation of agricultural assets, for the production of raw material, and for the development of mineral resources. In their train there follow, with the advent of industrial activity, the mechanical and electrical engineers. Their tasks are to make the fullest use of the revolution in ideas of transport, including transport by air, which have resulted from the perfecting of the internal combustion engine, and to secure the maximum advantage possible from cheap production and efficient distribution of electrical power. The day must come, to give a concrete instance, when the Victoria Falls, with their immense water resources, will mean much more for Africa than Niagara to-day means for America. Later still there will be called in the services of the chemical engineer, ever engaged in problems of research to ascertain the most advantageous processes of converting raw materials into manufactured articles. In all these tasks it is the South African engineer who has, under the conditions of an undeveloped land, built up a technique and practice suitable to African requirements and showing promise of wider applicability, that we may well expect to assume a position of leadership and of inspiration. These are some of the ways in which Science can respond to the challenge of Africa.

The picture which I set out to portray I have now completed. I have tried to suggest something of the magnitude of the rewards which Africa has in store for the scientist who has the enterprise to adventure and the vision to see. I have sought also to be the medium of the challenge presented to Science by Africa's opportunities and needs. It is a vast canvas on which I have had to work. On it I have drawn but a few sketchy outlines. Yet I hope that the vision stands out clear. I hope that I have said enough to convey the power of its inspiration. Not least do I hope that you, our visitors, will play a great part, in the time that you will spend with us, in filling in some of the details of the picture, and in quickening and vitalizing its message for the scientists of South Africa. It is to them chiefly that it makes its appeal. The development of Science in Africa, of Africa by Science, that is the Promised Land that beckons them. I believe that they will not be disobedient to the vision.

What invention of modern times is likely to have the greatest effect on human life ? Here we have a subject for endless discussion, but radio is certainly worth consideration in the debate.

No reference to radio would be complete without some mention of Marconi. It is true that his interests were concerned with wireless telegraphy rather than broadcasting, but the ether waves are the same : of very brief duration in the former case, and continuous in the latter. Maxwell told us all about these electro-magnetic waves as long ago as 1865—all about them, that is, except how to produce and detect them. Hertz first did this in 1887, but to Hertz all he had done was to confirm rather abstruse mathematical theory. Other investigators took up research on the waves from a purely scientific aspect, but Marconi, right from the first, seemed to sense their great importance, and he set himself the task of developing a practical system of communication by their means. He improved the coherer which was first used to detect the waves ; he first used the aerial and earth system, and he made the first experiments on short waves with a wire reflector behind the transmitting aerial. He returned again to the longer waves which promised more immediate progress, and by the beginning of 1901 he could signal for 200 miles. Then he adopted Lodge's ideas for improving the tuning of his circuits and set out to bridge the Atlantic, which he did in December. Within a few years of his success the greater number of ocean-going ships were fitted with wireless. But it seemed an awful waste to Marconi to send waves in every direction when one only wanted to send them between two points. *How was Beam Wireless developed ?* Marconi gives us the answer to this question.

G. MARCONI

RADIO COMMUNICATIONS

I HAVE always attached considerable importance to the problem of a practical directive system of radio communication. During my earliest experiments carried out in England more than twenty-eight years ago, I was able to show the transmission and reception of intelligible signals over a distance of one and three-quarter miles by means of an elementary beam system employing very short waves and reflectors, whilst, curiously enough, by means of the antenna or elevated wire system, utilizing much longer waves, I could only at that time get results over a distance of half a mile.

The progress made with the non-directional long-wave system was, however, so rapid and the results so immediately applicable to practical purposes that it very soon became, and still remains, what might be called the standard system. It is regrettable that the study of short waves was neglected for a long period of years, for these waves, which, so far, are the only ones that can be confined to narrow beams, are also capable of being employed to give practical results unobtainable by the lower frequency system, which, up to now, has held the field for long-distance communication.

When, during the War in 1916, I took up the systematic study of short waves, considerable doubt existed in my mind, and in that of other workers, as to whether the range of these waves might not prove to be too small for practical and useful purposes, particularly during day-time, if they might not be altogether too untrustworthy, and also as to whether large stretches of land, and particularly mountains, would not present obstacles to their transmission over long distances.

In 1920 experiments were carried out by Captain H. J. Round with duplex telephony on a 100-metre wave between Chelmsford and Southend, and the experiments were so successful that early in 1921 two stations which had been erected at Southwold and Zandvoord on the Dutch coast were put into commission experimentally, Southwold station utilizing about 1 kilowatt to the aerial. Experiments were carried out, transmitting from these two stations to Norway in August 1921, and at Christiansund day and night telephony was easily received from both stations. At Christiania, about 700 English miles distant, very loud and constant signals were received during the hours of darkness and in the day-time on certain days, apparently when the barometer was low.

During these experiments the curious night distortion of telephone signals was discovered, particularly when transmission was overland, the major cause of which has more recently been discovered by Captain Round in his work on broadcasting. Later, the results of these tests were merged into the general short-wave beam experiments.

During the tests carried out on the steam yacht *Elettra* in the spring and summer of 1923, I was able to discover that the short wave I was then using could not only cover great distances by day, and much greater distances by night, but that it was also quite trustworthy and that, moreover, large parts of continents and ranges of mountains did not materially reduce its working distance.

A series of tests was for the first time carried out with short waves over what might be termed world-wide distances during the winter, spring, and summer of this year, between Poldhu in Cornwall and receiving stations situated on ships at sea and also at such places as Montreal, New York, Rio de Janeiro, Buenos Ayres, and Sydney (New South Wales, Australia). All these tests proved to be successful, including the first telephonic communication with Australia ever realized, although the amount of power utilized at the sending station never exceeded 20 kilowatts. Very strong signals were obtained at all these places during the hours when darkness extended over the whole distance separating each of them from Poldhu, and weaker signals for a few hours when the sun was above the horizon at either end, the intensity of the signals varying inversely in proportion to the mean altitude of the sun when above the horizon. Although the signals were received with great strength at New York, Rio, and Buenos Ayres during the time when darkness spread over the whole or at least the major part of the great circle track separating these places from Poldhu, no signals at all were ever received, during these tests, when the same track or part of space was all, or substantially all, illuminated by the light of the sun.

While this limitation of the period of working to practically the hours of darkness constituted an undoubted disadvantage, still the economical advantages, together with the trustworthiness and possibility of working this system at far greater speeds than would have been feasible with the well-known high-powered long-wave installations, went far to convince me that the short-wave beam system would be capable of transmitting a far greater number of words per 24 hours between England and far-distant countries, such as Australia, than would be possible by the comparatively powerful, cumbersome, and expensive stations actually in use, or which were planned to be used, for Imperial commercial communications. It is a satisfaction to me to be able to state that the stations intended for this purpose in England and others to be installed in the principal Dominions and far-distant countries will all be on the beam system.

Commencing in August of this year, a further series of investigations was carried out between Poldhu and the yacht *Elettra*, the object being to endeavour, if possible, to find means of overcoming the limitations of working hours brought about by daylight, and also to test whether the effect of the reflectors would give the expected increase of signal strength over long distances. The yacht proceeded to Spain, then to Madeira, and afterwards to Italy. From Naples we sailed for Beirut in Syria, touching at Messina and Crete, returning to Naples via Athens.

At Madeira it was ascertained that a reflector at the transmitting station increased the strength of the received signals in accordance with our calculations, but that, notwithstanding this increase of strength, when using a 92-metre wave, the daylight range was only very slightly augmented. Comparative tests were carefully carried out with waves of 92, 60, 47, and 32 metres also at other places in the Atlantic and Mediterranean.

These tests enabled us to discover that the daylight range of practical communication over long distances increased very rapidly as the wave-length was reduced, the 32-metre wave being regularly received all day at Beirut, while the 92-metre wave failed to reveal itself for many hours each day even at Madeira, notwithstanding the fact that the distance between

Poldhu and Madeira is 1100 miles, entirely over sea, whilst that between Poldhu and Beirut is 2100 miles, practically all over mountainous land. Our observations went to confirm the fact that for waves between 100 metres and 32 metres the daylight absorption decreased very rapidly with the shortening of wave-length.

These results were so interesting and satisfactory that I immediately decided to try further tests over much greater distances. In October of this year, transmission experiments were carried out on a 32-metre wave from Poldhu to specially installed receivers at Montreal, New York, Rio, Buenos Ayres, and Sydney (Australia). Although the available power utilized at Poldhu was only 12 kilowatts, it was at once found possible to transmit signals and messages to New York, Rio, and Buenos Ayres when the whole of the great circle track separating these places from Poldhu was exposed to daylight. During a complete day transmission at fixed intervals carried out last October with Sydney, New South Wales, that station received the Poldhu signals for $23\frac{1}{2}$ hours out of the 24, and a 48-hour test which was only completed on December 10 fully confirmed the result.

The tests from England to places situated south of the equator, such as Sydney, Buenos Ayres, Rio de Janeiro, and Capetown, are particularly interesting for the reason that the waves have always in these cases to traverse what may be called a summer zone. They are therefore subjected to an averaging effect of conditions, which can never possibly exist when the transmissions take place only between stations situated in the northern or southern hemispheres. During November some successful receiving tests were carried out in England from a lower-power transmitting station in Australia utilizing waves of 87 metres. During the present month of December, trials have been continued with Canada, the United States, Brazil, the Argentine, and Australia, and also, for the first time, with Bombay and Karachi in India, and Capetown in South Africa. The power utilized at the Poldhu station during all these tests was 15 kilowatts.

The results have fully confirmed my expectations in regard to the behaviour of the various wave-lengths over such great distances, and I have no doubt that the information gained will render possible the installation of comparatively low-power stations capable of establishing and maintaining commercial service by day and by night between England and the most distant parts of the globe.

The low costs of this system both in capital and running expenses, compared with that of the existing type of stations, must prove to be very great, and should bring about the possibility of a reduction in telegraph rates for all long-distance communications, besides making direct communication with some of the smaller outposts of the Empire commercially remunerative. Already the size and the power of some of the most modern long-wave stations were becoming a serious question from a financial point of view. The newly equipped station at Buenos Ayres, for example, which was designed primarily for communicating with Europe over a distance oɪ about 6000 miles, employs 800 kilowatts and an aerial supported by ten towers, each 680 feet high. This station usually works on wave-lengths of about 12,000 and 16,000 metres. Another example is the British Post Office station which is being erected near Rugby, which, when completed, will employ 1000 kilowatts and an aerial supported by 16 towers, each 820 feet high, while the station being erected in the Union of South Africa was designed on a similar gigantic scale.

I am now firmly convinced that the beam stations employing only a small fraction of this power and much lower and fewer masts will be able to communicate at practically any time with any part of the Empire, and I cannot refrain from expressing my strong personal opinion that these powerful long-wave stations will soon be found to be uneconomical and comparatively inefficient in so far as long-distance commercial communications are concerned. Although we have, or believe we have, all the necessary data for the generation, radiation, and reception of electrical waves, as at present utilized for long-distance communications, we are still far from possessing anything approaching an exact knowledge of the conditions governing the propagation of these waves through space. These results indicate quite definitely that the well-known Austin formula is inapplicable to these waves. Another formula will have to be devised, based on the results of further investigations.

Reflectors of practical and economical dimensions are only efficiently applicable when short waves are used, and, although very long distances have been covered by these waves without the use of directional devices, I am convinced that these will be found to be essential for ensuring the carrying out of commercial high-speed services. The disadvantageous effect called " fading " is sometimes a source of serious trouble when receiving signals transmitted by means of short waves, although much less serious than when waves of several hundred metres in length are employed. According to our experience, the use of reflectors diminishes fading and also tends to overcome its effects by enormously increasing the strength and therefore the margin of readability of the received signals.

Increasingly large and expensive reflectors could, of course, be used with longer waves than 100 metres, but the results of all recent tests seem to indicate that the shorter waves present the greatest advantages, one of the most important being that their reception is very much less liable to interference by the effects of atmospheric electrical disturbances, or " X's ".

If these waves are destined to carry a considerable part of the most important long-distance telegraphic traffic of the world, it may well be necessary in the near future, by international legislation, to regulate their use and safeguard them from preventable interrerence.

A most important development of radio is radar, or radiolocation as it was first called. *How does Radar work?* For the answer to this question we turn to Sir Edward Appleton, whose pioneer research on the reflection of radio waves enabled radar to be developed in time to play a decisive part in repelling the German air-attacks on Britain. There are undoubtedly many people still alive who would have been victims of the Luftwaffe had it not been for radar. Sir Edward has often stated his belief that the scientist and the community ought to make joint decisions on the direction in which science should seek to make further progress. " The proper direction of scientific effort and the proper result of such efforts " he says " is one of the most important challenges of our time."

SIR EDWARD APPLETON

HOW RADAR WORKS

THERE must have been many people during the war who have asked the simple question : What is radiolocation, and how does it work ? And I am quite certain that, unless our war effort was helped by their being told, their question has not been answered. For the secret of radiolocation has been securely kept for ten years, and only now is it possible to disclose something about the way it works and what can be done with it. Now I am not going to start by giving you an elaborate and scientific definition of radiolocation. I am going to turn matters upside-down, and tell you first what radiolocation is not ; for I think it is best approached in that way. If an enemy aircraft or ship approached our coast, we should want to spot it as soon, and as far away, as possible. If the enemy aircraft or ship chose to send out radio messages while it was on its way, we should be able to detect it and find its position with radio direction-finders. That is an old technique, as old, almost, as radio itself. But it is not radiolocation as understood to-day. For in preparing to defend ourselves before the war, we knew that it was quite certain that the enemy would not announce his approach by radio in this way. We knew he would keep his radio quiet. A method had therefore to be found to spot him and get his position which would work whether or not he sent out radio signals. Such a method is provided by radiolocation. It does not require any radio assistance from the enemy. The only co-operation required from him is that of merely being there—existing as a solid body. For a solid body reflects radio waves. It stops them, and sends them back. It cannot help doing this. And it is by way of the waves it sends back that it gives away its position and can be detected. This process of radio reflection is therefore the basic feature of radiolocation, and the technical problem to be solved is to find out whether there is a reflection or not, and then, if there is, exactly where the reflection takes place.

Now the method of detecting radio reflections is an old one, twenty years old in fact, and it was first devised in purely scientific work, conducted with no thought of its present widespread practical application. Twenty years ago there was a bit of a squabble on in scientific circles concerning the existence of the Heaviside layer in the higher atmosphere. Some people believed that it existed and helped radio waves to travel round the earth. Others denied its existence. The problem was to settle it by direct experiment, by shooting radio waves up to it and seeing whether they were reflected and came back. The experiment was tried, both here and in America, and it was found that the waves did come back. But more than that, it was found possible in both sets of experiments to time the waves on their up and down journey, and, since we know how fast they travel, to calculate the distance they had travelled and so find where the Heaviside layer was situated—sixty miles above the ground, as it turned out. Now how was that done?

I expect that many of you have often stood a fairish distance away from a cliff, clapped your hands, and then listened for the echo. You know, of course, that the echo you hear is due to the sound which has travelled to the cliff, been reflected there, and then come back to you. If you could time the echo after the clap, you could find your distance away from the cliff. For example, if the echo came back in half a second, you would know that the cliff was about 250 feet away. Now the same method of echo-timing was used in the early radio measurements on finding the distance away of the reflecting Heaviside layer. But of course radio waves travel a million times as fast as sound waves, so rather special methods had to be used to measure the echo delay time. Still, it was done.

Now when you clap and listen for the echo from a cliff, you have got to keep quiet after you have clapped, in order to hear the echo. And that is just what you have to do in the radio case. The radio transmitter has to be made to send out a very short burst or " bang " of radio waves and then keep off the air for a while, so that the echo can be received clearly, and its delay measured. The timing of the echo is done with what is called a cathode-ray oscillograph —the kind of thing you have seen on your television set. In it there is a rapidly moving pencil of electrons which strokes across a glass screen from left to right making a bright line. As this pencil sweeps across the screen horizontally it gets a vertical kick when the burst of waves is sent out and, a bit later, another kick when the echo comes back. And from the separation of the kicks we can measure the delay time. But there is another important thing about the radio echo method and it is this. We have for many years known how to find the direction in which wireless waves arrive at a receiving station, and it is possible to do our direction-finding on the echo alone. This is rather easy, as a matter of fact, since the echo is shown on the cathode-ray tube-face as a separate signal on which we can concentrate our attention. So we get two things from our echo experiment : first, the range of the reflecting surface or body, and second, the direction in which it is situated. And if you think of that for a moment—direction and distance away—you will readily see that the position is completely specified.

Now so far I have dealt only with scientific experiments

and the methods used in them. How did it come about that these methods were adapted to practical ends in the radio-location of aircraft and ships ? An aircraft looks rather a tiny thing when seen in the sky and it is certainly a much smaller reflecting surface than the Heaviside layer. Who first gave us the hint that radio reflections from aircraft were detectable ? The answer is, I think, the engineers of the British Post Office, who were making some tests in 1931 using the short wave-length of five metres. They found that they could detect the presence of aircraft, by the wobbling of their signals, up to a distance of two and a half miles from their receiver. Tele-vision viewers, a little later, used to notice a similar kind of effect on their screens when aircraft were about. So you see that by 1933 we already knew all about the scientific principles of locating reflecting surfaces by radio, and we also had some idea of the reflecting power of aircraft. It remained to develop, on the basis of this knowledge, a military weapon which would give us ample warning of the approach of hostile aircraft. This work was undertaken in the first instance by a small group of the radio staff of the National Physical Laboratory, working under Sir Robert Watson-Watt, at a secret Air Ministry station on the east coast.

The importance of radiolocation was quickly grasped by each of the three Defence Services and soon there were either attached to, or working in collaboration with this station, scientific representatives of each Service, developing radio-location for their particular requirements. A little later these workers were joined by scientists from the universities and from industry. The whole radiolocation effort became, in fact, that of one large team. I stress all this because I want to make it clear that our successful development of this subject was due to the ideas and inventions of hundreds of people whose names you may never hear of. And don't forget this too : the experimental work of the scientist is not, by itself, enough. The equipment has to be manufactured before the Services can use it. There is, therefore, another vital chapter in the radiolocation story—that of radio production, in which hundreds of thousands of British workers played their all-important parts.

Well now, how did this great effort develop? You have already seen that the first thing we need to do in radiolocation is to illuminate the target with radio waves as strongly as possible. Now we can do this in two ways. If you go up into your loft to look for a trunk, you may either be able to switch on an electric light and illuminate the whole place, or you may have to take an electric torch and search round with it. There is the same kind of choice in radio nowadays, but there was not in 1935, for we were not then expert in producing the very shortest of waves, and to get a really sharp beam of radio waves you have got to have very short waves. The trouble with longer waves is that they spread. Thus the very earliest radiolocation stations had to use a fairly broad beam of illumination because they used longish waves. These, our first operational radiolocation stations, sent out their spurts of radio waves which bounced back, revealing the position of an aircraft, even though it was only half-way from Germany across the North Sea.

However, you will easily see that this floodlighting of a large area is wasteful in radio power. So it was not long before there was a move towards shorter and shorter waves with attempts to focus them with mirrors, as is done in a searchlight. This was, in fact, the second stage in the development of radiolocation and it was a revolutionary stage indeed. Instead of using wave-lengths which had been measured in metres, the new range of wave-lengths was measured in centimetres and so an entirely new radio technique was developed. This brought with it many advantages. First, since the radio beam could be kept clear of the ground, it got rid of unwanted echoes from permanent objects on the landscape. Secondly, it got over the difficulty of detecting low-flying aircraft coming in from the sea—a weakness which had shown up with the longer wave coast stations. Thirdly, since the aerial system used was small, the centimetre sets were most convenient for carrying on aircraft. And fourthly, when carried on aircraft, these sets were found particularly useful for showing up small targets such as submarines.

But there was still something more that could be done, and that takes us to the third stage in the development of radiolocation. The short-wave beam was eventually used to give a sort of television picture on a screen, a rough picture of what the landscape looks like as seen from an aircraft, or rather what it would look like to a pilot who could see in the dark, or could see through cloud. Anything that sticks up well above the surrounding territory always gives a good radio echo, so if the radio beam is made to "look at" or "scan" the landscape point by point, high buildings give responses, while flat portions, like streets and ponds, do not. By making the radio beam swing sideways and also backwards and forwards in a short time, the various echoes could be registered and pieced together on a screen that has a good memory to form a rough picture of the landscape. And when still smaller wave-lengths are available we may expect to get radiolocation television of still finer detail.

Now what does all this sum up to in weapons of war? There have been so many developments in radiolocation that it is impossible to mention them all, but here are a few. First, the word "search" in "searchlight" is now a misnomer, for searchlights can open up straight on the target. Anti-aircraft gunners can fire at an aircraft located above cloud through which no searchlight could possibly penetrate. Our own aircraft can be directed to meet the enemy from ground locators. When within range they themselves can detect the enemy with their own sets. They can also be given warning of an enemy on their tail. Ships can fire at an enemy hidden in a smoke-screen, or spot icebergs hidden in fog, and I have already mentioned the spotting of submarines and the detection of towns from the air.

I am one of the people who don't like to hear it said that

this is a scientists' war. The scientist may make the tools, but the fighting man uses them. The truth is that the war has given the scientist a unique opportunity of putting his special skill and knowledge to support the Fighting Services. Outstanding in this support has been the development of radiolocation—the main offering and the main tribute of British radio to our victorious armed forces.

Another modern invention of untold possibilities for the future is the electron microscope. The value of the ordinary microscope has long been realized, and the best instruments have a magnification of some 4000 diameters, which is equivalent to making a postage stamp look as big as a football ground. But a much more powerful tool is now in our hands. *How can electrons be used to give a picture of things far too small to be seen in the most powerful microscopes?* Sir Charles Darwin, who first satisfied a long-felt mathematical want by producing a set of equations which *did* correspond with the movements of an electron in an electro-magnetic field, gives us the answer. His lecture was given in 1943, and already much further progress has been made, more especially, perhaps, in preparing the " specimens " for examination. One way of doing this is by showering them with metallic atoms falling at an angle. In this way every little projection casts a shadow which is free of these metallic atoms, and which therefore allows the electrons to pass easily through it, whereas they are stopped by the layer of metal elsewhere. The electron microscope can magnify up to 100,000 times, which is equivalent to giving a halfpenny a diameter of a mile and a half. This is going to be of immense help in a great variety of ways, especially in medicine and in industry.

SIR CHARLES DARWIN

REVEALING THE INVISIBLE

I AM going to tell something about a new instrument that promises to be very useful in all sorts of ways. This is the electron microscope, and its great virtue is that it can

magnify things something like fifty times as much as an
ordinary one. This makes it reveal a lot of things that were
quite invisible before. For example, there were many diseases
which were known to be due to microbes, but the microbes
were too small to see in the best existing microscope, whereas
now we can photograph them.

When you think of a microscope you probably think of a
man sitting at a table in front of the window squinting with
one eye through a little brass tube. The electron microscope
is nothing like that. It stands about 9 feet from the floor :
the working part is in a column about a foot in diameter and
there are various parts of its gear alongside so that altogether
it covers about three feet square on the ground. The object
you want to study is put on a very little thin sheet of celluloid
in a small chamber about 6 feet from the ground, and at about
knee height there are a number of small windows at the sides
of the column that you can look down through to see the
image formed on a green luminescent screen.

Now, as to how it is done. The essential things in this
microscope, as in an ordinary microscope, are the lenses, but
the lenses are quite different here. To make an enlarged
image of anything the essential thing is to be able to bend
the rays coming from it. If it is possible to bend the rays
in any way at all, then by being ingenious in arranging how
they are bent it is possible to make an enlarged image of some
sort, and by very skilful designing to make a really good
enlarged image. When the rays are rays of light the way
this is done is by glass lenses, as in a telescope or a camera,
and you will appreciate what I said about skilful designing
when I tell you that it could easily take a team of highly
trained men a year to work out the shapes of the little lenses
—often eight of them one after the other—that go to make
up a new pattern of high-class microscope. But in the electron
microscopic we are not using light, and so the lenses have to
be quite different. We are making a beam of electrons, which
are the ultimate stuff of electricity, tiny particles which go to

make up a great part of matter. They were discovered by J. J. Thomson, and it is only because they are working in the valves of your wireless sets that you can listen to broadcast sound. The electrons are given off by a heated wire. In a radio valve they go comparatively slowly but in the microscope they are speeded up to about a quarter of the speed of light, which means that if you could make a pipe right round the earth and shoot them through it, they would come round and hit you in the back in about half a second. At this high speed the electrons can go right through a very thin sheet of celluloid but will be stopped or scattered by any rather thicker object such as a microbe that you put on the celluloid. Here we have a set of rays which can give a picture of the object if we can bend them. Electrons can be bent either by electric fields or by magnetism, and so we have either electric or magnetic lenses. Both work, but on the whole the magnetic ones have been more used. The lens consists of a circular coil of wire in a specially designed iron frame. It carries electric current, and this curls up the tracks of the electrons in a rather complicated way that has just the same property as a glass lens has for light, of bringing them to a focus.

We have now got the two requisites for a microscope, light (only here it is not light) and lenses, and we must put them together to make the microscope. The electrons start from a heated wire at the top of the microscope column and are speeded up so as to travel downwards. There are three lenses in the microscope. The first is called the condenser, and it concentrates the beam of electrons on the object you are examining. In an ordinary microscope there has to be a condenser lens too—though people often forget about it—to concentrate the light on to the microscope slide ; otherwise there would not be enough light to see by. The electrons now come to the object, say a microbe on a thin sheet of celluloid. The ones that hit the microbe are stopped, but the rest go on. Next, just below, there is another magnetic lens which could produce an image about 18 inches below, magnified about 100 times. For the small things you want to look at this is still much too small, so you do not form the image there but instead put another lens below which magnifies it again 100 times. So the final image which is looked at or photographed is about ten thousand times magnified, and as this is often still a bit small, you enlarge the photograph you have taken perhaps five or ten times more by ordinary photography. There is one peculiarity in the image of a magnetic lens that I might mention. In a camera or, for the matter of that, in your own eyes too, the image is exactly upside-down, and so it is in some telescopes, though in others a lens is put in specially to make it right way up, but with a magnetic lens it is twisted round so as to point in some other direction. You have to focus of course, which you do by altering the current in the lens, and as you turn the handle to do this you see the image slowly twisting round as it gets focused.

Now why is this new elaborate gear better than the old microscope? The answer is that in the old microscope, though you can magnify the image indefinitely you gain nothing by it beyond a certain point, as you see only a large blurred image instead of a small blurred image. There is a definite limitation to the size of the things you can see, which is the wave-length of the light you use. The wave-length of light is about a fifty-thousandth of an inch, and if you try to magnify up anything as small as this you cannot help losing all the sharp edges and you get only indefinite blobs. By very hard work, using ultra-violet light, something rather better can be done, but, roughly speaking, it is not worth magnifying more than about four thousand times. There is a similar theoretical limitation for electrons, but it only comes in for sizes many thousands of times smaller, and if it were the only limitation, we could hope to be just about able to see individual atoms. However, there is another limitation which is not so fundamental, but which threatens to be practically more serious, and this is that it seems unlikely that anyone will succeed in making a good enough lens to go beyond about a hundred times smaller than an ordinary microscope can do. At present most of the successful photographs are magnified not much beyond twenty thousand, though I have seen some very good ones at a hundred thousand.

As to what the microscope can be used for, there is the trouble that the object has to be in a vacuum, and so it is much easier to work with dry things. Particles of smoke of various kinds can be seen; some smokes are stringy, some are in little cubes and some in needles. Another trick is to study the roughnesses of an apparently smooth surface of metal or anything else, by coating it with a thin layer of resin which you afterwards peel off and put in the microscope. But probably the chief interest is in microbes and such things. Many of these can be seen with an ordinary microscope, but we now know that some of them, that looked like a blob, really had a swimming tail. There were many diseases known to be caused by microbes which were too small to be seen at all, but now we can see them. I have seen one beautiful photograph made in Germany, which shows the mysterious thing called bacteriophage, a beast which attacks and kills bacteria. You can see a large black object, the bacteria, and a crowd of things like tadpoles round it swimming towards it and attacking it. Without the electron microscope they were simply not seen at all. We have also ourselves made photographs of those other mysterious things called viruses which cause certain kinds of diseases. Altogether I think we can be pretty sure that in quite a few years a great deal more will be known about disease.

I expect you want to know who invented this remarkable instrument, but I have got to disappoint you. It is the thing everyone asks about any new invention, and it is nearly always impossible to answer. Inventions are hardly ever like that. It does not matter whether it is the telegraph or photography or wireless or radiolocation, you cannot say who was the inventor, because it is a gradual process of one man seeing something but not how to use it, another pushing it on a little way, and so on. Of all the really big inventions of the last hundred years there is only one where I should care definitely to name the undisputed discoverer, and that one is Röntgen's discovery of X-rays, and they were found more by luck than by judgment. The electron microscope is not in that class. The theory behind it was quite well known to physicists a long time ago, and the main question was to see that there could be practical results. I remember myself not so many years ago hearing about the business and thinking it was a pretty game but that nothing would come of it. Perhaps the earliest practical work was done by a number of Germans ten years ago or so, but quite a number of other workers were on to it, and laboratory instruments had been made and used in several countries both here and abroad. Now the thing is being manufactured and can be bought in America and we are fortunate in having half a dozen of their instruments in this country. They are being used for a great variety of purposes, and whatever may come of the work there can be no doubt that we shall within a very short time know a great deal more about what goes on in the world at the size of a millionth of an inch.

We have traced the growth of science, and taken a glimpse at just a few of the almost innumerable fields which it has covered. Perhaps the question has already suggested itself to many minds : *Will there ever be an end to scientific discovery ?* Will men ever be able to say that there are no more questions left unanswered, no more worlds left to conquer ? Perhaps a little caution may be advisable in dealing with this problem, so we give the answer in the form of a parable—extracts from " The Story of the Other Side " in Richard Jefferies' *Bevis*, that lovely account of an idealized summer in the life of a boy.

RICHARD JEFFERIES

THE STORY OF THE OTHER SIDE

" ONCE upon a time " said Bevis, closing his eyes now, " there was a great traveller who went sailing all round every sea—— "

" Except the New Sea " said Mark.

" Yes, except the New Sea which we found, and went riding over all the lands and countries, and climbing up all the mountains, and tramping through all the forests, and shooting the elephants and Indians and sticking pigs, and skinning boa-constrictors, and finding magicians ; and he went spying everywhere, and learned everything, and—— "

" Go on—what next ? "

" He went on till he said it was all no good, because if you went into the biggest forest that ever was you walked through it in about three years—— "

" Like they did through Africa ? "

" Just like it ; and if you climbed up a mountain, after a day or two you got to the top ; and if you sailed across the sea, if it was the greatest sea there ever was, you came to the other side in six months or so ; so that it did not matter what you did, there was always an end to it."

" Very stupid."

" Very stupid, very ; and he got tired of it always coming to the other side. He did so hate the other side, and he used to dawdle through the forests and lose his way, and he used to pull down the sails and let the ship go anyhow, and never touch the helm. But it was no use : he always dawdled through the forests after a while, and—— "

" The wind always took the ship somewhere."

" Yes, to the hateful other side, and he got so miserable, and what to do he did not know, and he could not stop still very well—nobody can stop still—and that's why people have got a way of spinning on their heels in some countries, I forget their names—— "

" Dervishes ? "

" Dervishes of course ; well, he became a Dervish, and used to spin round and round furiously ; but you know a top always runs down, and so he got to the other side again."

" Stupid."

" Awful stupid ; and so he said that this was such a little world he hated it. You could go all round the earth and come back to yourself and meet yourself in your own house at home in no time."

" It's not very big, is it ? " said Mark. " Nothing is very big that you could go round like that."

" No, and the quicker you get round the smaller it is, though it's thousands and thousands of miles, so he said ; and so he set out again to find a place where he could wander and never get to the other side, and after he had walked across Persia and Khorasan and Beloochistan——"

" And Afghanistan ? "

" Yes, and crossed the Indus and Ganges, and been over the Himalayas, and inquired at every temple and of all the wise men who live in caves and hang themselves up with hooks stuck through their backs——"

" Fakirs."

" At last a very old man took pity on him, seeing how miserable he was, and whispered to him where to go, and so he went on——"

" Where ? "

" To Thibet."

" But nobody is allowed to enter Thibet."

" No ; but he had the password, which the aged man whispered to him, and so they let him come in, and then he wandered about again for a long while, and by this time he was getting very old himself and could not walk so fast, so that it took longer and longer to get to the other side each time. Till at last, inquiring at all the temples as he went, they promised to show him a forest to which there was no other side. But he had to bathe and be purified first, and they burned incense and did a lot of magical things——"

" In circles ? "

" I suppose so. And then one night in the darkness, so that he should not see which way they went, they led him along, and in the morning he was in a very narrow valley with a wall across so that you could not go any farther down the valley, nor could you climb up, because the rocks were so steep. Now, when they came to the wall he saw a little narrow bronze door in it—very low and very narrow—and the door was all covered with carvings and curious inscriptions——"

" Magic ? "

" Yes, very magic. And the man who showed it to him, and who wore a crimson robe, over which his white beard flowed nearly down to the ground—I am sure that is right, flowed nearly down to the ground, that is just what my grandpa said—the old man went to the door and spoke to it in some language he did not understand, and a voice answered, and then he saw the door open a little way, just a chink. Then he had to go on his hands and knees, and press his head and neck through the chink between the bronze door and the wall, and he could see over the country which had no other side to it. Though you may wander straight on for a thousand years, or ten thousand years, you can never get to the other side but you always go on, and go on, and go on——"

" And what was it like ? "

" Well, the air was so clear that he was certain he could see over at least a hundred miles of the plain, just as you can see over twenty miles of sea from the top of a cliff. But this was not a cliff, it was a level plain, and he could see at least a hundred miles. Now, behind him he had left the sun shining brightly, and he could feel the hot sunshine on his back, but inside the wall there was no sun——"

" No sun ? "

" No. Ever so far away, hung up as our sun looks hung up like a lamp when you are on the hill by Jack's house—ever so

far away, and not so very high up, there was an opal star. It
was a very large star and so bright that you could see the beams
of light shooting out from it, but so soft and gentle and
pleasant that you could look straight at it without hurting your
eyes, and see the flashes change exactly like an opal—a beautiful
great opal star. All the air seemed full of the soft light from
the star, so that the trees and plants and the ground even
seemed to float in it, just like an island seems to float in the
water when it is very still, and there was no shadow——"

"No shadow?"

"No. Nothing cast any shadow, because the light
came all round everything, and he put his hand out into it
and it did not cast any shadow, but instead his hand looked
transparent, and as if there was a light underneath it——"

"Go on."

"From the bronze door there was a footpath leading out,
out, winding a little, but always out and out, and so clear was
the air that though it was only a footpath, he could trace it for
nearly half the hundred miles he could see. The footpath was
strewn with leaves fallen from the trees, oval-pointed leaves ;
some were crimson, and some were gold, and some were black,
and all had marks on them.

"One of these was lying close to the bronze door, and as he
had put his hand through, as you know, he stretched himself
and reached it, and when he held it up the light of the opal
sun came through it—it was transparent—and he could see
words written on it which he read, and they told him the secret
of the tree from which it had fallen.

"Now, all these leaves that were strewn on the footpath
each of them had a secret written on it—a magic secret about
the trees, and the plants, and the birds, and the stars, and the
opal sun—every one had a secret on it, and you might go on
first picking up one and then another, till you had travelled a
hundred miles, and then another hundred miles, a thousand
years, or ten thousand years, and there was always a fresh
secret and a fresh leaf.

" Or you might sit down under one of the trees whose branches came to the ground like the weeping ash at home, or you might climb up into another—but no matter how, if you took hold of the leaves and turned them aside, so that the light of the opal sun came through, you could read a magic secret on every one, and it would take you fifty years to read one tree. Some of the leaves strewed the footpath, and some lay on the grass, and some floated on the water, but they did not decay, and the one he held in his hand went throb, throb, like the pulse in your wrist.

" And from secret to secret you might wander, always a new secret, till you went beyond the horizon, and then there was another horizon, and after that another, and you could go on and on, and on, and though you could walk for ever without weariness because the air was so pure and delicious, still you could never, never, get to the other side."

BIOGRAPHICAL NOTES

E. N. DA C. ANDRADE, F.R.S. (1887-   ), head of the Royal Institution (1950-52) and a leading physicist of our age, is regarded with equal admiration by scientists for his original researches and by a thoughtful public for his eminently readable books on popular science.

He was born in London, and after a most distinguished University career, including post-graduate work in London, Heidelberg, and the Cavendish Laboratory, he interrupted his scientific work in 1914 to serve with the R.G.A. He was appointed Professor of Physics in the Artillery College, Woolwich, in 1920, and Quain Professor of Physics in the University of London in 1928. He took over the Chair at the Royal Institution and command of the Davy-Faraday Research Laboratory in January 1950. He used smoke to enable him to photograph sound waves in the air ; he made metal wires consisting of a single crystal ; he collaborated with Rutherford in the work which established the nature of the gamma-rays of radium ; and he has carried out systematic research on phosphorescence and luminescence —light without heat.

His main publications are *The Structure of the Atom* among his technical works, and *Engines* ; *The Mechanism of Nature* ; *Simple Science* (with Julian Huxley) ; *The New Chemistry* ; and *The Atom and its Energy* among popular books. He also wrote an authoritative life of *Isaac Newton*.

SIR EDWARD APPLETON, F.R.S., G.B.E. (1892-   ), Principal and Vice-Chancellor of Edinburgh University, has made his mark both as a scientist and as an administrator. It was his work on the reflection of radio waves which proved the existence, first of the Heaviside layer some 60 miles up, and then of the Appleton layer at twice that height. It was this work, also, which established the principle of radar.

Appleton was born and spent his schooldays at Bradford. He then went to St. John's College, Cambridge, where he carried all before him in the Natural Science Tripos. In the first World War he became a wireless officer in the Royal Engineers, and when hostilities ended he went to the Cavendish Laboratory to begin his research on wireless " fading ". From 1924 to 1936 he was Wheatstone Professor of Physics in London University, and then Jacksonian Professor at Cambridge until 1939. In that year he became Secretary of the Department of Scientific and Industrial Research, a post which he held for ten arduous but very fruitful years. He was awarded the Nobel Prize for Physics in 1947. His administrative experience was summed up in the shrewd remark, " It is possible to get almost anything done, if one is prepared *not* to claim credit for it "

ARISTOTLE (384-322 B.C.), tutor of Alexander the Great, and founder of the first " Lyceum " must be included in our survey of science, although his greatest work was probably done in politics, ethics, and philosophy. Even in science his main contributions were to biology. He was born at Stagira in Macedonia, the son of a wealthy physician, and at the age of 17 he was sent to Athens to study in Plato's Academy. He stayed there for twenty years—indeed until Plato's death—when he went to Asia Minor and married. Soon afterwards he became tutor to the young prince Alexander.

In 336 B.C. Philip was murdered, Alexander succeeded to the throne of Macedonia, and Aristotle's services were no longer needed. He returned to Athens, and, disappointed at not being elected leader of the Academy, founded his own rival school at the Lyceum. The next twelve years were the most fruitful of his life. They were devoted to research and to discussion and writing—work which was to dominate the scientific thought of the western world for eighteen centuries. Aristotle quarrelled with Alexander over the latter's attempts to make the Macedonians pay him divine honours, but in spite of this the Athenians regarded him as an enemy when they rose against Macedon, and Aristotle had to flee to Chalcis where he spent the last year of his life.

E. F. ARMSTRONG (1878-1945) was Scientific Adviser to the Ministry of Home Security in the second World War. Science was in his blood. He was the son of the famous H. E. Armstrong who used to make even the austere pages of *Nature* sparkle with his wit, and his own son, K. F., had already shown brilliant promise when an unfortunate Alpine accident ended his life.

Edward Frankland Armstrong specialized in the chemistry of sugars, and in physical chemistry, but he was not satisfied with academic research in the laboratories. He wanted to see discoveries applied in British industry, and he held a number of important directorships in large firms. He was elected F.R.S. in 1920, was President of the Society of Chemical Industry from 1922 to 1924, and gave long and valuable service to the Royal Society of Arts, first as Treasurer, then as President and Chairman of the Council.

F. W. ASTON, F.R.S. (1877-1945), scientist and craftsman, invented the mass-spectrograph by means of which he was able to separate the atoms of different weight in the same substance—an achievement of amazing delicacy, and of immense importance, since the making of the atom bomb depends on the separation, on an engineering scale, of the 235 isotope of uranium from the more common, but comparatively inert, 238 isotope.

Francis William Aston was born in Birmingham, and studied science at Mason College, where he spent most of his spare time acquiring a remarkable skill in glass-blowing. In 1910 he joined Sir J. J. Thomson's team at the Cavendish Laboratory, and there, except for a war-time interlude when " work of national importance " superseded work of international importance, he remained. He was elected a Fellow of the Royal Society in 1921, won the Nobel Prize for Chemistry in the following year, and was awarded a Royal Society medal in 1938. In his prime he was a splendid athlete, and a great player of tennis, golf, and bridge.

SIR ROBERT BALL (1840-1913), the famous astronomer, was born in Dublin and educated at Trinity College. He served his apprenticeship as an astronomer in the Earl of Rosse's observatory at Parsonstown, and while there discovered four spiral nebulae. On the death of Lord Rosse he became Professor of Mathematics in Dublin University, and in 1874 was appointed Royal Astronomer of Ireland, a post he held until 1898. He was knighted in 1886, and elected Lowndean Professor of Astronomy and Geometry at Cambridge, and Director of the Observatory in 1892. He was President of the Royal Astronomical Society from 1897 to 1899.

Of his books the best are *The Story of the Heavens, Great Astronomers* ; and a *Popular Guide to the Heavens*. His more learned works include a *Treatise on Spherical Astronomy* and a *Treatise on the Theory of Screws*.

DR. J. BRONOWSKI (1908-   ), Director of the Central Research Establishment of the National Coal Board, was born in Poland and rivals his compatriot Conrad in his command of the English language. After graduating at Cambridge he went to University College, Hull, as Lecturer in Mathematics, and in 1942 was seconded to Government Service as head of a series of mathematical and statistical units planning the economic destruction of the Axis powers. He was scientific deputy in the Chiefs of Staff Mission to Japan which reported on the atomic bomb damage. On his return he was engaged in applying statistical research to the economics of industry until his appointment in 1950 to the Coal Board Research Centre.

His first broadcast was given on the night of the Bikini explosion, and dealt with the atomic bombs in Japan. It was well entitled " Mankind at the Crossroads ". He has often broadcast on scientific subjects, as well as taking part in such diverse programmes as the Brains Trust and " Any Questions ? " His main personal interest is the relation between the arts and science, and his publications include *The Poet's Defence* ; *William Blake : a Man without a Mask* ; and *The Common Sense of Science*. He is the author of many mathematical and scientific papers.

NICOLAS COPERNICUS (1473-1543). The man who revived the heliocentric theory of Aristarchus and " put the earth in its place " was born in Thorn on the Vistula, and after learning Latin and Greek at home, was sent to the University of Cracow to study medicine. He qualified as a doctor, but he found that his main interest was in mathematics and natural science. After his graduation he went to Italy, and soon made a reputation, being appointed Professor of Mathematics in the University of Rome.

His uncle, however, had become Bishop of Ermeland, and being very proud of his nephew he asked him to return home and take up a canonry at the cathedral of Frauenberg. Copernicus accepted the offer and made a leisurely way back, studying in Padua for a time. He soon settled down at Frauenberg, for he was a peace-loving man, and his great interest in life was astronomy. He wanted to study the stars, and nothing was allowed to stand in his way. He had no transit instrument, so he knocked slits in the walls of his house, set up a quadrant in front of them, and was thus able to measure the times and the elevations of the stars and planets as they crossed the meridian. Soon he began to take a special interest in the motions of the planets. Mars proved particularly puzzling. Its brightness and its magnitude varied so much that not even Ptolemy's epicyclic motion would explain the contrasts. Gradually he became convinced that the only explanation was that the sun and not the earth was at the centre of the universe. Both Earth and Mars went round the sun, and that was why the brightness of Mars varied so much. Sometimes it was near the earth, sometimes very far away.

Copernicus stated and proved his theory in his book, *The Revolutions of the Celestial Orbs*, which was practically finished by about 1530. But he was a good churchman as well as an astronomer. He knew what an upsetting effect it would have, and it was many years before his friends and admirers persuaded him to publish it. At the age of 70 he had a stroke, and the printed book was placed in his hands just a few hours before he died.

JOHN DALTON (1766-1844) the inventor of " atomic weights " was born near Cockermouth of Quaker parents, and developed a taste for mathematics at an early age. After twelve years as a schoolmaster in Kendal he became Professor of Mathematics and Natural Philosophy in New College, Manchester. In 1799 he became Secretary of the Manchester Literary and Philosophical Society. He was a keen student of the weather, and having discovered that he was colour-blind he wrote the first scientific paper on that disability. But his most important work is his atomic theory which gives a quantitative explanation of all chemical combinations. On this, modern chemistry has been built up.

SIR CHARLES G. DARWIN, F.R.S. (1887-   ) was Director of the National Physical Laboratory from 1938 to 1949, and is now a Member of the Museums and Galleries Commission. His work on quantum mechanics earned him the Royal Medal of the Royal Society in 1935. He was the first mathematician to produce equations which *did* correspond to the observed movements in an electro-magnetic field of that incredibly small, but incredibly complicated " wavicle " (wave plus particle) the electron.

He was born in 1887 and educated at Marlborough and Trinity College, Cambridge. From 1910 until 1914 he was Reader in Physics at Manchester University, and then he enlisted and served in the R.E. and R.F.C., being promoted to Captain and winning the M.C. At the end of the war he became Lecturer in Mathematics in Christ's College, Cambridge, and in 1923 he went to Edinburgh as Tait Professor of Natural Philosophy. He has published a large number of papers on theoretical physics, *The New Conceptions of Matter* (1931), and *The Next Million Years.* He was knighted in 1942.

SIR ARTHUR EDDINGTON (1882-1944) was the leading British exponent of the theory of relativity when Einstein first launched this on the scientific world. Born at Kendal, he was educated at Owen's College, Manchester, and Trinity College, Cambridge, where he was Senior Wrangler in 1904 and Smith's Prizeman. After serving for some years as chief assistant at the Royal Observatory at Greenwich he was appointed Professor of Astronomy at Cambridge, and Director of the observatory there. Many honours came his way : F.R.S., the

Gold Medal of the Royal Astronomical Society, a knighthood and the Order of Merit.

His main researches, apart from relativity, dealt with the motions and evolution of the stars, and he published many highly technical papers and books on these subjects. Among his more popular works the best-known are *Stars and Atoms* and *New Pathways in Science*.

MICHAEL FARADAY (1791-1867), whose researches made possible the dynamo and the electric motor, was earning his living as an errand-boy at the age of 10. Later he was apprenticed to a bookbinder, and, by great good luck, an occasional binder of scientific books. The insides of the books interested Faraday much more than the bindings. Although his family was very poor he began to make experiments for himself, and eventually, thanks to Sir Humphry Davy's sympathetic reaction to a letter he wrote, he was appointed laboratory assistant at the Royal Institution in 1813. After many years of brilliant work on a variety of problems, mainly chemical, he succeeded, during the autumn of 1831, in proving his theory of electro-magnetic induction, which is the principle of the dynamo, and the foundation of our electrical age. As a chemist of no mean order, Faraday was also responsible for pioneer research work into what is now known as electro-plating. He was always true to his own youthful vision of the philosopher as a man " willing to listen to every suggestion, but determined to judge for himself. He should not be biased by appearances, have no favourite hypotheses, be of no school, and in doctrine have no master. He should not be a respecter of persons, but of things. Truth should be his primary object." Faraday was elected F.R.S. in 1823, and very soon he became widely known as a practical scientist of exceptional ability. He could have made a fortune as a consultant to industrialists. He prefered his fundamental research work at the British Institution, and he died a poor man.

GALILEO GALILEI (1564-1642), one of the greatest of all names in the history of science, was born at Pisa in Italy. His father was an impoverished Florentine nobleman who at first wanted to make Galileo into a cloth merchant. But the boy had much too vivacious a mind for such a calling; he was obviously meant for the university, and so his father consented to his studying medicine. Like Copernicus, however, Galileo turned from medicine to mathematics and science, and at the age of 26 he was appointed Professor of Mathematics at Pisa. There it was that he made his famous experiment in dropping bodies of different weights from the Leaning Tower, and there, too, he studied the problem more closely by measuring the times taken by brass balls to roll down an inclined plane. He had no clock so he measured the time by letting water leak from a pail while the ball was in motion, and then weighing the amount that had escaped.

His bold criticism of Aristotelian mechanics made him many enemies at Pisa and he had to resign his Chair. But he was soon offered the Professorship of Mathematics at Padua, and there he began to uphold the Copernican system and to speak strongly against the Ptolemaic. He made his telescopes, and his fame grew rapidly. He was the idol of the Venetian Republic where thought was, for those days, remarkably free.

In 1610, however, he moved to Florence, and from that time onwards he was under the watchful eye of the Church, receiving a strong warning in 1616 not to teach his heretical opinions. When Urban VIII became Pope, however, in 1623, Galileo thought that he could safely go back to his Copernican teaching again. The new Pope was one of his friends, and a tolerant and enlightened man. So Galileo published his *Dialogues on the Ptolemaic and Copernican Systems* with the well-known disastrous results.

He claimed that his Dialogue was impartial, but the Aristotelian arguments were given by a character called Simplicio who certainly lived down to his name, while the Copernican champion, Salviati, always argued keenly and well. It was, in fact, amazing that the Master of the Sacred Palace ever gave permission for the book to be published. It created an uproar, was quickly banned, and Galileo was summoned to Rome, where he was made to recant. His last years were spent in what has been well described as " a modified exile ". It was sweetened by the homage of eminent men from all over Europe.

SIR ARCHIBALD GEIKIE (1835-1924), the Scottish geologist, was one of those rare scientists who combine profound knowledge with a gift for exposition and for making his subject appear both simple and interesting. He was born at Edinburgh, and educated at the High School and University of his native city. He showed an early bent towards geology, and joined the staff of the Geological Survey in 1855, helping to produce the geological map of Scotland which was published in 1862. In 1865 appeared his *Scenery of Scotland* which dealt with his favourite topic of how the scenery had been made. In 1881 he was appointed head of the Geological Survey, a post he held for ten years. He travelled and observed extensively through Europe and

America, and published many scientific volumes. He was President of the Royal Society in 1909 and received the Order of Merit in 1914. His *Landscape in History and other Essays* is perhaps the most interesting of his many publications.

H. F. VON HELMHOLTZ (1821-94), the inventor of the ophthalmoscope, was a descendant of the Quaker, William Penn. He was born at Potsdam, and studied medicine, serving as a surgeon in the Prussian army. He was appointed Professor of Physiology at Königsberg in 1849, and held similar Chairs at Bonn and Heidelberg in later years. He became Professor of Physics at Berlin in 1871, and Director of the Physicotechnical Institution of Charlottenburg in 1887.

Helmholtz made many important advances in the theories of hearing and of vision—more particularly of colour-vision—and, with Lord Kelvin, he was one of the founders of the theory of the conservation of energy.

JAN H. HOFMEYR (1894-1948), scholar and statesman, was probably the most brilliant of South Africa's sons. He was educated at the South Africa College, and at Balliol, Oxford, where he went as Rhodes Scholar and took a double first. He returned to the University of the Witwatersrand, first as Professor of Classics, then as Principal, then Vice-Chancellor, and finally as Chancellor. In 1924 he turned to politics as Administrator of the Transvaal. In 1933 when General Hertzog formed a " best man " Cabinet for his Coalition Government, Hofmeyr was given the Portfolio of the Interior, which included Education and Health. He resigned in 1938 in protest against the appointment of Mr. Fourie as Native Senator ; and his enemies, and even some of his friends, said that his career was finished and that he was too much of an idealist for these realistic days. Twelve months later came the war, and " Hoffie " was soon back again, second in command to General Smuts, and destined, apparently, for the greatest of all political honours. His budgets were remarkably successful, but a most untimely death ended his career. His publications include a biography of his grandfather *Onze Jan*, *Studies in Ancient Imperialism*, *The Open Horizon*, and *South Africa* in the Modern World Series.

RICHARD JEFFERIES (1848-87) is the British naturalist who succeeded, perhaps better than any other writer, in giving expression to his deep sense of communion with Nature. He was born at Coate near Swindon, the son of a farmer who was not a successful business man. After attending school until he was 15 he tried unsuccessfully to work his way first to Russia and then to America. He returned home and worked as a local journalist for a time, but in 1872 he sent to *The Times* a remarkable letter on " The Wiltshire Labourer ". This revealed his true gifts, both to himself and to others, and he began to write on farming and farmers for *Fraser's Magazine*. A number of delightful books were written in the brief remaining years, notably *The Game-keeper at Home*, *Wild Life in a Southern County*, *Wood Magic*, and *Life of the Fields*. But perhaps the best of all is the transmuted auto-biography of a single summer in his boyhood, *Bevis*, from which the parable that closes our volume on Science is taken.

SIR HAROLD SPENCER JONES, K.B.E., F.R.S. (1890-   ), Astro-nomer Royal from 1933 to 1955, worthily carried on the helpful traditions of Greenwich Observatory, founded in 1675 for the advance-ment of navigation through progress in astronomy. In 1939 he installed at Greenwich a new type of quartz clock which keeps even better time than the earth itself.

Born in Kensington, he was educated at Latymer Upper School, Hammersmith, and Jesus College, Cambridge, where he took a Triple First in Mathematics and Natural Science. He was Chief Assistant at the Royal Observatory for ten years from 1913, spent the next ten years as H.M. Astronomer at the Cape of Good Hope, and then returned to Greenwich. Ten years later, to keep up the regular cycle, he was given the honour of knighthood, and the Gold Medals of the Royal Society and the Royal Astronomical Society. But many other honours and degrees came in the years between.

He has written many books, both popular and specialist. Of the former, his *Worlds without End* and his *Life on Other Worlds* are perhaps those best calculated to arouse an appetite for more.

IMMANUEL KANT (1724-1804), the great German philosopher, was born at Königsberg, the son of a saddler of Scottish descent. He lived a completely regular and uneventful life (he never married) and was Professor of Logic and Metaphysics at the Königsberg University from 1770 until 1797. His life was spent in an endeavour to reconcile the conflicting realist and idealist schools of philosophical thought. See the volume on Philosophy and Ethics for further details.

A. L. LAVOISIER (1743-94), founder of modern chemistry, and victim of the Terror in the French Revolution, was born in Paris and educated at the Mazarin College. He was elected a member of the Academy of Sciences in 1768 and Director of the Powder Works in 1776. He rose to be Director of the Academy of Sciences in 1785. His interests were very wide, but he is remembered chiefly for his work in chemistry. He it was who first explained burning as an active process of combination with the oxygen in the air, and having come to this revolutionary but correct conclusion he was able to apply quantitative methods not only to his experiments on oxidation, but also to all his other work in chemistry. Thus he established the all-important principle of the conservation of matter, and put into general use the modern idea of an " element " as opposed to the old Greek conception. Lavoisier was secretary of the commission on weights and measures whose work led to the foundation of the metric system. He did much more excellent public work, but he was unfortunately one of the tax-gatherers, the *fermiers généraux*, and in 1794 the Convention ordered the arrest and trial of all those relics of the *ancien régime*. He was sent to the guillotine to show that " the Republic has no need of scientists ".

SIR CHARLES LYELL (1797-1875), the " father of modern geology ", was born in Forfarshire, the son of a well-known botanist. His education at Exeter College, Oxford, and at Lincoln's Inn led him to the bar, but he had become interested in geology while he was at Oxford, and he communicated his first paper to the Geological Society in 1822. He made a geological tour in Scotland in 1824, was elected F.R.S. in 1826, and in the following year he abandoned the law and henceforth devoted all his energies to science.

His main work on *The Principles of Geology* was already taking shape in his mind, and during the next few years he specialized in the study of the Tertiary formations. His classification of these into Eocene (Dawn of Recent), Miocene (Less of Recent), and Pliocene (More of Recent) has long been in common use. His *Principles* came out in 1830-33, followed by the *Elements of Geology* and, much later in 1863, *The Antiquity of Man*. All his books stress the fact that the forces which produced geological changes in the past are still at work ; that they can be seen at work ; and that they are slowly shaping the future lie of the land.

DR. HECTOR MACPHERSON (1888-1956), minister of the Guthrie Memorial Church in Edinburgh, was educated privately, proceeding to the University of Edinburgh and to New College, Edinburgh, where he became Waterbeck Prizeman and Cunningham Fellow. He specialized in history and astronomy, and was Elder Lecturer in Astronomy at the Royal Technical College of Glasgow. He has written a large number of popular books on Astronomy, and his

Guide to the Stars serves as an excellent introduction for anyone beginning the study of this subject. Perhaps the best of his historical books is *The Covenanters under Persecution*, while his devotion to religion and to science is well illustrated in *The Church and Science*.

MARCHESE MARCONI (1874-1937) did more than anybody else to make wireless communications possible. He was born at Bologna in Italy, the son of an Italian country gentleman and an Irish mother. At an early age he showed a keen interest in physics, and especially in electricity. His first experiments in wireless were made at his father's villa, and even with his own crude home-made apparatus he succeeded in sending messages for a distance of a mile. Then he came to England where better facilities for experiment were available. That was in 1896, and two years later he had established regular communication over the 14½ miles between Alum Bay and Bournemouth. His distances quickly increased. In 1899 his system was tried in the British naval manœuvres, and by the beginning of 1901 he could cover 200 miles. On December 12, 1901, his signals were heard across the Atlantic. Progress after that was swift and world-wide.

Marconi received many honours and awards : an Honorary D.Sc., from Oxford in 1907, the Nobel Prize for Physics in 1909, the G.C.V.O. in 1914, and in 1929 he was created Marchese.

JAMES CLERK MAXWELL (1831-79), planner and first Director of the Cavendish Laboratory at Cambridge will always be remembered for his full mathematical treatment of electro-magnetic waves. His pioneer work has guided all workers since his time in the study of wireless waves, heat rays, light, X-rays, and all other forms of radiation As Professor Niels Bohr put it, " Maxwell's theory, besides being extremely fruitful in the interpretation of phenomena, has yielded the utmost any theory can do, namely, to be instrumental in suggesting and guiding new developments beyond its original scope ".

He was born in Edinburgh, and educated at the Academy and the University of his native city. After studying at Cambridge he spent a few years as Professor of Natural Philosophy at Aberdeen, and then came to London to take the Chair of Physics and Astronomy at King's College. In 1871 he was called to the newly founded Chair of Experimental Physics at Cambridge, and the Cavendish Laboratory was opened three years later. Students were few at first, but Maxwell infected them with his own zeal and enthusiasm, and the Laboratory soon built for itself a tradition of successful discovery. For five years Maxwell, always accompanied by his dog, controlled and inspired the work at Cambridge, and then, all too soon, death cut short his brilliant career.

SIR JOHN L. MYRES (1869-1954) had a number of honorary degrees awarded to him. The Universities of Wales and of Manchester gave him degrees in Science ; Athens gave him a degree in Philosophy ; and the Witwatersrand gave him one in Literature. Perhaps South Africa made the best choice, although all the honours were most richly deserved.

Myres was born at Preston in Lancashire and educated at Winchester and New College, Oxford, where he took a double first. He was Student and Tutor at Christ Church from 1895 to 1907, and Lecturer in Classical Archaeology from 1903 to 1907. Then he moved to Liverpool University as Professor of Greek, and from 1910 until 1939 he was Wykeham Professor of Ancient History. His long and distinguished academic career was interrupted by the first World War, when he served as Acting Commander in the R.N.V.R. He travelled much in Greece and Asia Minor and directed some important excavations in Cyprus.

His publications include *A History of Rome*, *The Dawn of History*, *The Political Ideas of the Greeks*, and *Who were the Greeks ?*

ISAAC NEWTON (1642-1727) is the " super-nova " among the stars of British science. His most famous remark : " I do not know what I may appear to the world ; but to myself I seem to have been only like a boy playing on the sea-shore and diverting myself now and then in finding a smoother pebble or a prettier shell than ordinary, whilst the great ocean of Truth lay all undiscovered before me " was made shortly before his death. It reflects very faithfully his own attitude to his famous discoveries in mathematics and science. He had invented the differential calculus ; he had produced a theory of light and carried out some wonderful experiments on refraction ; and he had formulated and given a mathematical proof of his theory of gravitation as the universal law which kept the stars and planets in their places. He was probably the greatest scientific genius that has ever lived, and yet, for him, his most outstanding discoveries were only pebbles and shells. The " great ocean of Truth " was mysticism and religion. This is characteristic of his attitude to science ; he could work at some problem with intense energy and brilliant concentration for a time, but when it was solved he lost interest in the subject and usually needed some outside stimulus—a challenge or even an insult from another scientist—

to spur him into renewed activity, especially in the latter half of his life.

Newton was born on Christmas Day, 1642, and heredity can give no explanation at all of his unusual powers. He went to the King's School at Grantham, and then to Cambridge University. After a year or two his mathematical genius seems to have flowered quite suddenly and he was elected a scholar of Trinity College. The plague drove him out of Cambridge in 1665, and while he was back in his native village of Woolsthorpe he took the first vital steps in his three great discoveries already mentioned. But he was never in a hurry to publish his findings, and all these were only given to the world later, and in piecemeal fashion.

In 1696 he was appointed Warden of the Mint, and for three years had a very busy time supervising the issue of a new coinage, and the withdrawal of the debased currency—a task which he performed with great efficiency. Soon after his appointment, and while his mind was doubtless fully occupied with his new duties, the mathematician John Bernoulli set two difficult problems and challenged anyone to solve them in six months. At the request of Leibnitz he extended this period by an additional year. Newton received the problems on one day and sent off both solutions on the next.

SIR WILLIAM OSLER (1849-1919) was one of mankind's benefactors who " never left a sick-room without leaving behind renewed hope ". He was born at Bond Head in Canada and studied at Trinity College, Toronto, and McGill University, Montreal, where he took his M.D. He continued his studies in London, Leipzig, and Vienna, and returned to McGill in 1874 as Professor of Medicine. He held Chairs in the States between 1884 and 1904, first at Pennsylvania and then at Johns Hopkins Universities. In 1905 he was appointed Regius Professor of Medicine at Oxford. He carried out much valuable research work, notably on the spleen and on the heart. Apart from his medical books, which were long regarded as standard works, he wrote a volume of essays, *Aequanimitas,* and *A Way of Life.*

KARL PEARSON (1857-1936) played a leading part in applying mathematics to the theories of evolution and heredity. He was born in London and educated at King's College, Cambridge. Called to the bar in 1882 he soon found the study of eugenics more interesting than the law, and he was appointed Galton Professor of Eugenics in the University of London, and Director of the Francis Galton Laboratory.

He was elected a Fellow of the Royal Society in 1896, and awarded the Society's Darwin Medal. His publications include *The Grammar of Science* which has been through three editions, *National Life from the Standpoint of Science*, and *The Life, Letters and Labours of Francis Galton*.

WILLIAM PEDDIE (1861-1946) was born at Westray, one of the Orkney Isles. He studied science at Edinburgh, and was elected F.R.S.E. in 1898, the year in which he won the Makdougall-Brisbane Medal. After lecturing at Edinburgh he was appointed to the Chair of Physics in University College, Dundee, in 1907, and he held this post until he retired in 1942. He wrote some standard monographs on physical science, notably *A Manual of Physics*, *Molecular Magnetism*, and *Elementary Dynamics of Solids and Fluids*. He played a leading part in the co-operative efforts of psychologists, physicists, and physiologists to solve the problems of vision and colour vision.

CECIL FRANK POWELL, F.R.S. (1903-   ) has been Melville Wills Professor of Physics in Bristol University since 1948. He was educated at Judd School, Tonbridge, and Sidney Sussex College, Cambridge, where he took a First in Natural Science. Appointed Research Assistant to Professor A. M. Tyndall at Bristol, he was a member of the expedition organized jointly by the Royal Society and the Colonial Office to study seismic and volcanic activity in the island of Montserrat in 1935. He was Vernon Boys Prizeman of the Physics Society in 1947. His researches in recent years, which have been on nuclear physics and on the use of photographic methods of investigating this difficult but vitally important subject, earned him the Nobel Prize for Physics in 1950. He is the author, with G. P. S. Occhialini of *Nuclear Physics in Photographs*.

PTOLEMY OF ALEXANDRIA (*c.* 100-178) flourished in the scientific revival which took place under the rule of the Emperor Marcus Aurelius. He collected the works of all his known predecessors in astronomy, and out of these plus his own observations he built up a complete astronomical system. Unfortunately he based it on Aristotelian principles. Lighter bodies move more slowly than heavier ones. So the earth must be fixed and immovable, otherwise it would long since have lost all its lighter substances such as the air. The heavenly bodies, being perfect, must move in circles round the earth, and to explain the awkward movements of the planets he said they moved in circles round a point (" the deferent ") which itself moved in a circle round the earth. Ptolemy built up a clever table of chords which served most of the purposes of the modern table of sines, and his trigonometrical knowledge helped him to support his theory with mathematical figures.

He was also a famous geographer, compiling the first maps with lines of latitude and longitude on them. He gave an account of all known countries from Norway to India and China.

PYTHAGORAS OF SAMOS (*c.* 570-497 B.C.) is best known as the discoverer of the famous theorem about the squares on the sides of a right-angled triangle, which he probably proved in a way just as ingenious as Euclid's. Little is known for certain about his life, but he migrated to Croton, a Greek Colony in Southern Italy about the year 529 B.C. Here he became the leader of a brotherhood which at first was religious rather than political, but which aroused some opposition from the authorities, and this led to Pythagoras retiring to Metapontium, where he stayed until his death.

His greatest discovery was that of the length-ratio of stretched strings necessary to produce musical intervals : 2 to 1 for the octave, 3 to 2 for the fifth, and 4 to 3 for the fourth. This perhaps led him to the theory that " all things are numbers ". The Pythagoreans used to swear by the " holy Tetractys " which was ten dots arranged in a triangle and showing at a glance that $10 = 1 + 2 + 3 + 4$. By another arrangement of dots Pythagoras showed that the sum of a sequence of odd numbers beginning with 1 is always the square of a number. $1 + 3 = 2$ squared. $1 + 3 + 5 = 3$ squared, and so on. Pythagoras was one of the first to believe that the earth and all the heavenly bodies are spheres. He did so because the sphere is the most " perfect " of solid figures.

LORD JOHN RUSSELL (1792-1878), a champion of all the great Whig and Liberal measures of his time, was born in London and educated at the University of Edinburgh. He entered Parliament at the age of 21, and began his advocacy of Parliamentary Reform in 1819. His liberal views found more and more support among the Whigs and he became leader of the party in 1834. He was Prime Minister from 1846 to 1852, and again in 1865. As Secretary of State for the Colonies in 1839 he claimed the whole of Australia for Britain, and added New Zealand to the empire as well. He accepted an earldom in 1861. His interest in science and his great debating skill have been inherited by his grandson the third Earl (Bertrand Russell).

LORD RUTHERFORD (1871-1937), is the man who "broke up the indivisible, made plain the invisible, changed the immutable and unscrewed the inscrutable "—all by splitting the nucleus of the atom, or at least by long years of hard and eminently successful work on that mysterious ultra-microscopic assemblage of powers and particles.

Ernest Rutherford was born at Nelson in New Zealand, and after graduating with high honours came to work under Thomson at the Cavendish Laboratory. He then went to Montreal as Professor of Physics in McGill University, and while he was there he established the fundamental theory of radio-activity, the exponential law. In 1907 he returned to England to occupy the Chair of Physics at Manchester, and in the following year he was awarded the Nobel Prize for Chemistry. He gave us his theory of the "solar system" type of atom in 1911, and this theory was expanded and developed by Niels Bohr. Honours came thick and fast: F.R.S., a knighthood, the Order of Merit, President of the British Association and of the Royal Society, and a peerage. He was Director of the Cavendish Laboratory from 1919 until his untimely death following an abdominal operation.

DR. ERWIN SCHRÖDINGER (1887-   ), Professor in the University of Vienna, is the leading exponent of quantum wave-mechanics, a most important aspect of the Quantum Theory. He was born and educated at Vienna, and after holding Chairs at Stuttgart, Breslau, and Berlin, he came to Magdalen College, Oxford, as a Fellow in 1933. From 1940 to 1956 he was Senior Professor in the Dublin Institute for Advanced Studies. With P. A. M. Dirac he shared the Nobel Prize for Physics in 1933 for his work on the Quantum Theory, and he was elected a Foreign Member of the Royal Society in 1949. His publications include a vast number of highly technical papers on quantum mechanics, and also *Science and the Human Temperament, What is Life?* and *Expanding Universes.*

SIR J. J. THOMSON (1856-1940), the discoverer of the electron, was a member of the Order of Merit. But to scientists all over the world he was G.O.M.—the Grand Old Master. From 1884 to 1918 he was head of the Cavendish Laboratory. His example and influence inspired hundreds of men who were his pupils, and through them thousands of others who were the pupils of his pupils.

Thomson was born near Manchester and educated at Owen's College, Manchester, and Trinity College, Cambridge. On taking his degree he began to work in the Cavendish Laboratory under Lord Rayleigh. Four years later in 1884 Rayleigh resigned, and Thomson, much to his own surprise, was appointed his successor. Never was an appointment more fully justified. Every year brought fresh confirmation of this fact until 1918 when he resigned to become Master of Trinity. Thomson was elected F.R.S. in 1884, won the Nobel Prize for Physics in 1906, was knighted in 1908, and was President of the Royal Society from 1916 to 1920. His *Discharge of Electricity through Gases,* and *Conduction of Electricity through Gases* are standard technical works.

BIBLIOGRAPHY

HERE are the titles of a few books which can be recommended for further reading by our Members :—

Karl Pearson. *The Grammar of Science* (Everyman).

C. J. Singer. *A Short History of Science to the Nineteenth Century.*
    [A general survey from Greek to modern times.]

H. Butterfield. *Origins of Modern Science.*
    [Covers the sixteenth, seventeenth, and eighteenth centuries.]

A. N. Whitehead. *Science and the Modern World.*

E. N. da C. Andrade. *The Mechanism of Nature.*

W. Bragg. *Old Trades and New Knowledge.*

A. S. Eddington. *New Pathways in Science.*

H. Spencer Jones. *Worlds Without End.*

J. Jeans. *The Stars in their Courses.*

Sherwood Taylor. *The World of Science.*

ACKNOWLEDGMENTS

For permission to include copyright material in this volume our acknowledgments and thanks are due to :—

Professor Andrade and Messrs. G. Bell & Sons for " The Philosopher and the Scientist ", and to Professor Andrade for " Science the Servant of Humanity " ; to Sir Edward Appleton for " How Radar Works " ; to the Royal Society of Arts for E. F. Armstrong's " Science in Nineteenth-Century Britain " ; to the *British Journal of Radiology* for F. W. Aston's " The Structure of the Atom " ; to Dr. J. Bronowski for " A Sense of the Future " and " Dr. Einstein Sums Up " ; to Sir Charles Darwin for " Revealing the Invisible " ; to the Cambridge University Press for Eddington's " Gravitation and Relativity " from *Space, Time and Gravitation*, for the extract from *Beyond the Electron* by J. J. Thomson, and for the translation of Galileo's " Discovery of Jupiter's Satellites " from *Cambridge Readings in the Literature of Science* ; to Macmillan & Co., for " The Value of Organized Knowledge "· taken from Geikie's *Landscape in History* ; to Sir Harold Spencer Jones for " Life on Other Worlds " ; to the Rev. Dr. MacPherson for " The Universe as Revealed by Modern Astronomy " ; to the Oxford University Press and Sir John Myres for " The Beginnings of Science " from *Science and Civilization* in the Unity of Civilization series ; to Professor E. S. Pearson and J. M. Dent & Sons for the extracts from *The Grammar of Science* by Karl Pearson ; to Professor C. F. Powell for " The Most Transient Form of Matter " ; to Dr. Schrödinger and *The Times* Publishing Co., for " Matter Dematerialised " from the *Progress of Science* ; and to the British Association for the Advancement of Science for a number of lectures which were originally delivered at the annual meetings.

Every effort has been made to trace the owners of copyrights, and indulgence is craved for any inadvertent omission.

INDEX

(Titles of Lectures are in Black Type)